The Power of Design: Al(Ga)N - A Material System for Advanced Device Performance

Aurora

Table of Contents

CHAPTER 1:

INTRODUCTION

1.1. BACKGROUND

III-Nitrides, a group of direct bandgap compound semiconductor material system that includes binary materials such as AlN, GaN, InN as well as their ternary and quarternary alloys, are of wide interest due to the large range of available bandgaps ranging from 0.7 eV (InN) and 6.2 eV (AlN)[1, 2]. The large range of available bandgaps makes these materials important for various applications ranging from optoelectronics, power electronics, and high frequency high power devices. One of the most prominent III-N material, GaN, has already started making a significant impact in the power and RF electronics arena [3]. In fact, GaN-based RF power amplifiers are currently overtaking Si laterally diffused metal oxide semiconductor (LDMOS) RF power amplifier market share[4]. GaN is making an impact in the high-performance power electronics market as well, due to the low on-resistance and high breakdown voltages for GaN devices with smaller dimensions compared to Si devices, enabled by the higher critical breakdown field[5].

While significant progress has been achieved with regards to GaN electronics over the past few decades, the need for the ever-increasing improvement in performance and power efficiency requires the community to look for material systems that can overcome the limitations of GaN. The key aim of this book is to look at one such potential material system, Al(Ga)N, elucidate and understand the key benefits of these materials and correspondingly design devices which can overcome the limitations in GaN.

1.2. POLARIZATION OF GaN AND AL(GA)N MATERIAL SYSTEM

Both GaN and Al(Ga)N material system display extremely large electronic polarization fields, which includes both spontaneous and piezoelectric polarization[2]. This polarization presents us with a unique device design approach where it is possible to introduce charges as high as 3×10^{13} cm^{-2} without the necessity of introducing any dopants in the system. This ensures that dopant related scattering is not an impediment in the operation of these devices. It is important to note that due to the uniaxial nature of the wurtzite crystal, the polarization fields exist along the c (0001) axis and is absent in the major axes on the c-plane, which is one of the key reasons why most of these materials are grown in this orientation.

The discontinuity of polarization at the AlGaN/GaN heterojunction has been used to induce a two-dimensional electron gas (2DEG) in a high electron mobility transistor (HEMT) which uses the 2DEG for channel charge conduction. In these devices, a thin layer of metal-face Al$_x$Ga$_{1-x}$N film (barrier) is epitaxially grown on a Ga-face GaN or Al$_x$Ga$_{1-x}$N film (channel), which causes a 2DEG to be formed at the heterojunction [6-8]. It is also possible to grow the AlGaN/GaN devices in the N-face [9]. There is no need for modulation doping for these HEMTs, as is required for AlGaAs/GaAs heterostructure HEMTs, as the charge is fully generated by the polarization effect. The 2DEG charge density is dependent on the thickness of the barrier layer and the Al-composition difference between the barrier and the channel.

1.3. HIGH FIELD TRANSPORT PROPERTIES IN ALGAN

Monte Carlo simulations have predicted the high field saturated velocity of electrons in GaN HEMTs to be as high as 3×10^7 cm/s while time delay analysis in scaled GaN HEMTs have shown the velocity to be in the range of 1-2.8×10^7 cm/s[10-13]. In general, for a scaled transistor, as the gate length is decreased, the electric field increases which leads to an increase in the average velocity of the carriers. Transistors typically have significantly non-uniform electric field profiles with peaks near the gate edge, and hence the velocity is defined as an average quantity. To evaluate the effects of scaling, simulations were performed using TCAD modeling software, Silvaco ALTAS[14]. The simulations (Figure 1.1) were performed assuming a low mobility material with a mobility of 100 cm^2/Vs, a saturated electron velocity of 1×10^7 cm/s and a sheet charge density of 1×10^{13} cm^{-2}. As can be seen, as the transistor dimensions are shrunk the average velocity increases and approaches the saturation velocity. This is an important strategy to design, high f_T RF transistors using low mobility materials.

Based on the polar local oscillator (LO) phonon-based for electron velocity in these 2DEG configurations it has been shown that the electron velocity varies as $\frac{1}{\sqrt{n_s}}$ [15]. For materials such as GaN and AlGaN, the saturation velocity is limited by the optical phonon energy. As the electric field is increased in these materials, the electrons in the 2DEG gain energy until they reach the optical phonon energy, at which point an optical phonon is emitted and the electron loses its energy. The saturation electron velocities in these materials are thus dependent on the optical phonon energy and the effective electron mass. As the Al-composition in AlGaN increases, the optical phonon energy increases from 92 meV in GaN to ~100 meV in AlN[16]. However, effective electron

mass in AlN in about twice that in GaN. The resultant effect is that the electron velocity

in high Al-composition is expected to be lower than GaN but should be comparable in

magnitude.

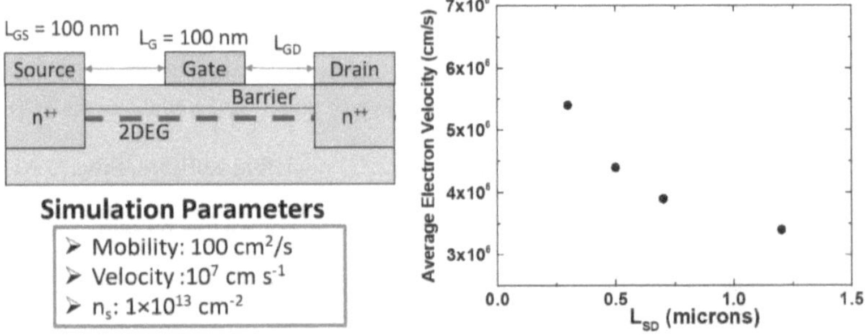

Figure 1.1 TCAD simulation showing that scaled transistors have higher average electron

velocity in the channel due an increase in the E-field intensity. The left figure shows the

schematic of the structure while the right figure shows the simulated average velocity as a

function of source to drain spacing.

Al(Ga)N exhibits high breakdown fields and good transport properties, which make them ideal candidates for high-speed and high-power amplifier applications. High Al-composition AlGaN, due to their ultra-wide band gap energy, have estimated breakdown fields much greater than GaN (Figure 1.2), which can be used to fabricate scaled devices with high sheet charge densities which can result in higher output powers at higher frequencies.

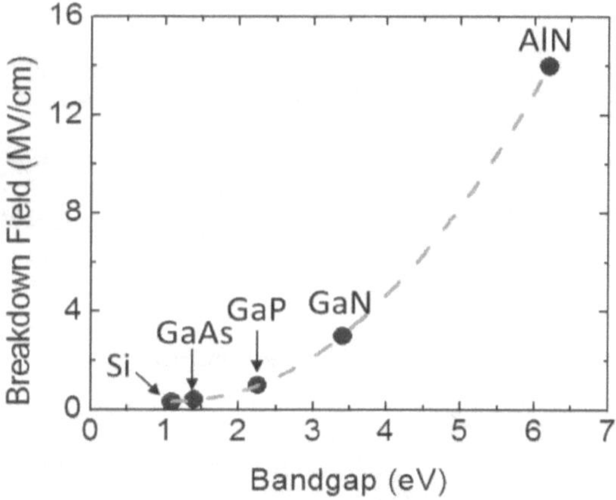

Figure 1.2 Breakdown fields of various semiconductors as a function of energy band gap. The value for AlN is extrapolated based on a square law dependence.

As shown previously by Johnson et al.[17], the product of the breakdown voltage (V_{BR}) and current gain cutoff frequency (f_T) are dependent on material parameters, and is given by

$$V_{BR} f_T = \frac{F_{BR} v_{sat}}{2\pi}$$

where F_{BR} and v_{sat} are the breakdown field and saturated velocity, respectively. The Johnson Figure of Merit and the analysis discussed later uses the cutoff frequency defined at the maximum drain bias and assumes a constant velocity. To estimate the transit time limited power density (P), we use the approximation for a Class A amplifier

$$P = \frac{I_{max} V_{BR}}{8}$$

where I_{max} is the maximum current density (A/mm), and V_{BR} is the breakdown voltage of the transistor.

Using the relationship between the cutoff frequency

$$f_T = \frac{v_{sat}}{2\pi L_{GD}}$$

and the maximum oscillation frequency f_{max}:

$$f_{max} = \frac{f_T}{2} \sqrt{\frac{R_0}{R_{in}}}$$

where R_0 and R_{in} are the output and input impedance, respectively.

7

Recognizing that the operating frequency is related to the maximum oscillation frequency by $f = f_{max}/\sqrt{G_P}$, where G_P is the power-gain, we can write the power density in terms of the operating frequency:

$$P_{out} = \frac{I_{max}\,F_{BR}\,v_{sat}}{32\sqrt{2}\pi W \sqrt{G_p}\,f}\sqrt{\frac{R_0}{R_{in}}}$$

where I_{max} is the maximum current density (A/mm), and $V_{BR} = \frac{1}{\sqrt{2}}F_{BR}\cdot L_{GD}$ is the breakdown voltage of the transistor[18]. Assuming an optimal channel charge given by $I_{MAX} = \frac{1}{\sqrt{2}}\cdot F_{BR}\cdot\epsilon_S\cdot v_{sat}$ we can represent the output power density as

$$P_{OUT} = \frac{\epsilon_S\cdot F_{BR}^2 v_{sat}^2}{64\pi\cdot\sqrt{G_p}\,f}\sqrt{\frac{R_0}{R_{in}}}$$

This relationship is plotted for GaN, $Al_{0.50}Ga_{0.50}N$ and $Al_{0.70}Ga_{0.30}N$ in Figure 1.3(a) as a function of operating frequency. Here, a thermal limit of 40 W/mm is assumed to be upper limit of maximum achievable power density. Figure 1.3(b) shows the output power as a function of the gain in dB for devices operating in the W band (94 GHz). The state-of-the-art results for GaN transistors are shown as well. As can be seen from both the plots, compared to GaN channel transistors, AlGaN channel transistors can enable orders of magnitude higher output power for the same operating frequency. Besides, that for operations near 100 GHz (Figure 1.3(b)), AlGaN channel transistors will provide substantially higher output power than a GaN channel transistor. This is because GaN is limited by intrinsic critical breakdown electric field and electron saturation velocities at these frequencies.

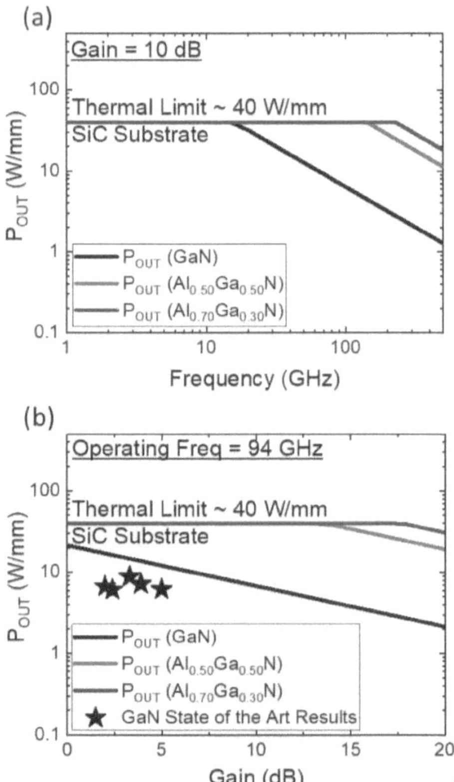

Figure 1.3 (a) Calculated power density as a function of operating frequency for gain = 10 dB for GaN, Al$_{0.50}$Ga$_{0.50}$N and Al$_{0.70}$Ga$_{0.30}$N. (b) The expected gain for GaN, Al$_{0.50}$Ga$_{0.50}$N and Al$_{0.70}$Ga$_{0.30}$N for an operating frequency of 94 GHz. GaN state of the art results are also shown in the figure[19-21].

1.5. book GOALS AND SUMMARY OF KEY RESULTS

The objective of this book is the design of ultrawide bandgap (UWBG) high performance aluminum gallium nitride ($Al_xGa_{1-x}N$, $x \geq 0.5$) channel devices for applications in high frequency, high power amplifiers. These devices, enabled by the high electron saturation velocity and the high critical breakdown electric fields in these materials, should in principle be capable of operation at high frequencies with significantly higher power density than incumbent technologies[22-26]. With the advent of the internet of things and the constant need for faster telecommunications across increasing number of interconnected devices, this work can potentially enhance the synergy between humans and technology by enabling ultra-efficient $Al_xGa_{1-x}N$ electronics.

Compared to the 6–8 GHz frequency of operation used for 4G networks[27], the operation frequencies of networks of the future will be greater than 30 GHz (mm-wave, 30–300 GHz). The incumbent GaN technology is limited by intrinsic breakdown electric field and velocity at such high frequencies and have output power density lower than 10 W/mm power near 100 GHz (Figure 1.3(b)). However, $Al_xGa_{1-x}N$ ($x \geq 0.5$) with saturated electron velocity comparable to GaN and much higher critical breakdown electric field, can potentially achieve 2–3× higher power density at mm-wave spectrum (Figure 1.4). Hence, AlGaN channel devices are expected to play an important role in enabling the next generation communications systems[28].

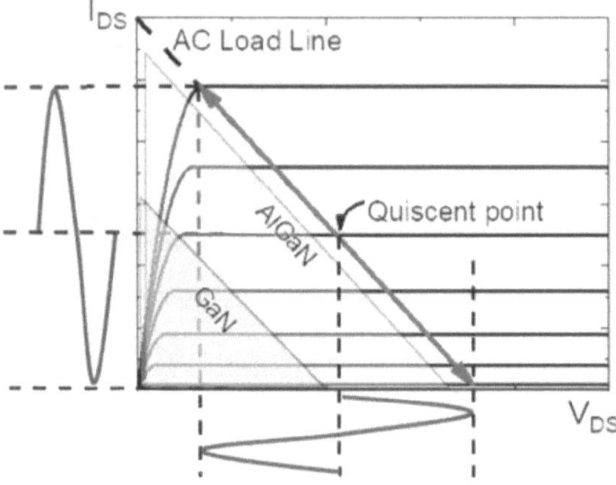

Figure 1.4 Illustration of the power triangle for AlGaN and GaN channel devices

Despite potential benefits, several challenges face the realization of high-performance AlGaN-based devices with superior breakdown characteristics. Firstly, ohmic contact formation is challenging due to the low electron affinity in these materials. One potential solution is the use of heterostructure engineered reverse Al-composition graded contact layers with heavy [Si$^+$] doping using Molecular Beam Epitaxy (MBE) growth technique. Currently, the highest quality AlGaN films can be grown by using a highly controlled ultra-high vacuum growth technique called metalorganic chemical vapor deposition (MOCVD) growth technique[29]. However, reverse Al-composition graded contact layer method has not successful in achieving low resistance contacts for MOCVD grown films mainly due to the low doping efficiency for MOCVD grown materials. The second challenge is the difficulty of effectively utilizing the high critical breakdown field available in these materials and approaching the material breakdown limit in actual lateral devices remain a standing challenge to this date.

While $Al_xGa_{1-x}N$ displays high F_{BR}, the drawbacks of having high contact resistance and high sheet resistance, due to their low mobility which is discussed in detail in Chapter 3, make realizing high performance RF devices a challenge. The feasibility of $Al_xGa_{1-x}N$ channel transistors for RF performance up to 100 GHz has already been demonstrated in the literature via simulation. The schematic and the electric field profile of the device are shown in Figure 1.5. In this study, using the test case of $Al_{0.75}Ga_{0.25}N$ channel and assuming an electron mobility of 100 cm^2/V·s, it has been demonstrated that for a scaled transistor with a gate length of 100 nm and a source-drain spacing of 800 nm, f_T can potentially exceed 100 GHz with a breakdown voltage of 100 V and current density of 2.5 A/mm[24]. The electron saturation velocity for this has been conservatively assumed to be 1×10^7 cm/s.

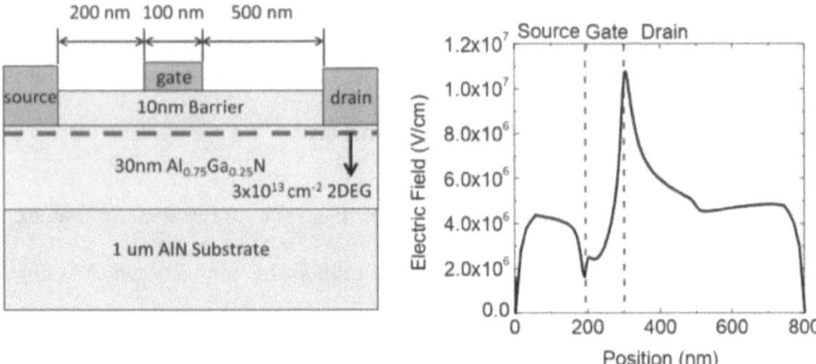

Figure 1.5 (left) Schematic of a 100 nm Al$_{0.75}$Ga$_{0.25}$N channel HEMT with a 10 nm barrier and (right) Electric field profile of the device at a gate to drain bias of 100 V (courtesy Dr. Zhanbo Xia)

In this book, to address the problem of in situ ohmic contact formation in MOCVD grown reverse graded contact layers, we have used novel polarization charge engineering of the reverse Al-composition graded contact layer to significantly reduce the contact resistance. In reverse graded contact layers, as Al-composition is reduced, bound negative polarization charges are formed which need to be compensated by using at least a commensurate concentration of [Si$^+$] dopants to produce low resistance electric current conduction. However, due to the low doping efficiency in MOCVD grown AlGaN films, high concentration [Si$^+$] doping is difficult. Focus was therefore shifted on the redesign of the contact layers instead to reduce the negative polarization charge created by reverse grading. This enabled low contact resistance even with low [Si$^+$] doping ($< 3\times10^{18}$ cm^{-3}).

No such studies have been done previously and this presented a suitable platform to understand the design of the contact layers in depth. Through this, a framework for the redesign of the reverse graded contact layers using an analytical model for the charge density in the contact layers and a self-consistent one-dimensional Poisson-Schrodinger solver[30]. With the redesigned MOCVD grown reverse graded contact layer, a contact resistance of 3.3×10^{-5} $\Omega\cdot cm^2$ was achieved to n-type $Al_{0.70}Ga_{0.30}N$ films. This is significantly lower than the previous lowest contact resistance achieved for $Al_{0.70}Ga_{0.30}N$, which was 3.5×10^{-4} $\Omega.cm^2$. A comparison of the specific contact resistivity for state-of-the-art MOCVD-grown $Al_xGa_{1-x}N$ channel transistors with $x > 0.5$ is shown in Figure 1.6. The specific contact resistivity obtained with the redesigned contact layers is the lowest observed for any MOCVD grown $Al_xGa_{1-x}N$ channel devices to date for $x > 0.5$. Metal semiconductor field effect transistors fabricated on these samples demonstrated current densities as high 635 mA/mm, which is the highest observed for any $Al_xGa_{1-x}N$ channel MESFET with $x > 0.5$. Another approach that was developed in this book was the hybrid growth of MBE regrown reverse Al-composition graded contact layers on MOCVD grown channel devices. A thorough study of the growth of such devices is presented in this book together with an experimental and modeling study of the regrown layer to understand the limitations and pave the path forward for further development of this technique.

Lateral devices break down significantly before the material limits are reached, and this is due to the premature breakdown of the gate-semiconductor junction, and the non-uniform electric field distribution between the gate/anode and drain/cathode in a transistor/diode. To solve this long-standing problem, a novel technique of integrating extreme permittivity materials, such as $BaTiO_3$, on the AlGaN devices was utilized. $BaTiO_3$ can be inserted between the gate/anode metal and semiconductor and between the gate/anode and drain/cathode regions where the electric field non-uniformity exists. This method can improve breakdown due to two reasons. Firstly, a high permittivity layer greatly reduces the field peaking for lateral devices and improves localized field peaking related breakdown. Secondly, the high permittivity material below the gate/anode metal reduces leakage which suppresses gate/anode leakage related breakdown.

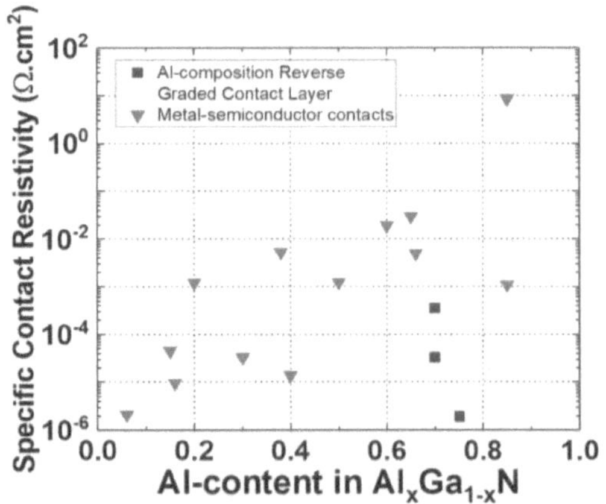

Figure 1.6 Comparison of the best reported contact resistances for different $Al_xGa_{1-x}N$ channel devices [29, 31-45]. **Note:** the trend of an exponential increase (y-axis is in log scale) in the contact resistance with increasing Al-composition for the metal-semiconductor contacts.

With the integration of the extreme dielectric constant material, significant improvement in the breakdown field was observed. Using 25 nm extreme dielectric constant material, $BaTiO_3$, on $Al_{0.58}Ga_{0.42}N$ heterojunction diodes, a 2–3× improvement in the breakdown performance was observed compared to the conventional Schottky diodes fabricated using the same $Al_{0.58}Ga_{0.42}N$ films. In general, for an anode to cathode spacing in the range of 0.2–0.3 μm, the average breakdown fields observed were in the range of 6–8 MV/cm. The highest measured breakdown field was 8.5 MV/cm, which is the highest experimentally observed average breakdown field for any semiconductor device to date.

The next 4 chapters in this book provide a detailed description of the various approaches that have been adopted to solve the contact resistance and premature breakdown improvement in high Al-compostion AlGaN channel devices, and state of the art results obtained by the adoption of these novel approaches.

Chapter 2 starts off with a discussion of the importance of contact resistance in achieving high performance RF and power devices followed by a literature survey of the traditional metal-semiconductor based approaches applied to the realization of the same for high Al-compositions AlGaN channel devices. The chapter then goes on to discuss the foibles of the traditional metal-semiconductor based contact formation schemes and establishes the importance of using a novel band-engineered heterostructure based contact formation scheme. This was realized using a reverse Al-composition graded contact layer inserted between the metal and the semiconductor. With the importance of the reverse graded contact layer established, the chapter furthers discussion of the various innovative contact formation schemes that have been applied in this book for the formation of low-resistance contacts. These include MBE-regrown contact layers on MOCVD grown channel where the contact was established for one case utilizing a top-regrowth approach and in another case using a side-regrowth approach. Details of the current limitations of these approaches are also provided both from an experimental and modeling perspective. The chapter also delves into a novel in situ slow-reverse graded contact layer which helped achieve record low contact resistance for any MOCVD Al-composition grown contact

17

layers with Al-content exceeding 50%. The techniques established in this chapter forms the essential backbone that enables the device results in the later chapters.

In chapter 3 the importance of mobility in a semiconductor transistor is discussed which is followed by an in-depth measurement and analysis of an AlGaN channel HEMT device. Using various scattering models, it is established that significant improvement in the mobility can be achieved in AlGaN by improving growth quality. Using this model further, an estimation is made about the upper limit of the mobility in AlGaN channel devices.

Chapter 4 talks about one of the most important problems in ultra-wide bandgap semiconductors whereby approaching the actual critical electric field breakdown limit in such materials, with breakdown electric field > 6 MV/cm, cannot be achieved in lateral devices. This is due to a) the non-uniform electric field distribution in the depletion region for a lateral device (which causes electric field peaking) and b) the bottleneck of tunneling related breakdowns at the Schottky gate or anode electrode. The chapter then discusses a novel solution where an extreme permittivity material, such as $BaTiO_3$, is inserted between the metal and semiconductor layers and between gate and drain region. The chapter discusses two devices, one an AlGaN channel diode and another an AlN/GaN transistor. Both devices show significantly improved breakdown characteristics.

The book finally discusses the effect of FinFET based device approaches on the contact resistance of AlGaN channel diodes. In this chapter, detailed discussion is provided on the process development of Fin-shaped AlGaN channel devices together with an analytical model to estimate the electric breakdown conditions of such devices. Using the analytical model, a comparison between the RF performance of planar AlGaN devices and FinFET AlGaN devices is also provided. This is followed by the demonstration of the improvement in current density and contact resistance on actual AlGaN channel transistors.

CHAPTER 2:

CONTACT FORMATION TO HIGH AL-CONTENT ALGAN

CHANNEL TRANSISTORS

2.1. BACKGROUND

As discussed in the introductory chapter, $Al_xGa_{1-x}N$, a ternary alloy of GaN and AlN, displays a broad spectrum of bandgaps from 3.4 eV (GaN) to 6.2 eV (AlN) and has various device applications ranging from optoelectronics to RF amplification and power switching applications[46]. These materials are, in fact, expected to have electron saturation velocity (v_{sat}) comparable to GaN and should show similar levels of current density. UWBG AlGaN is expected to have high critical breakdown field, F_{BR}, exceeding 10-11 MV/cm for $x > 0.7$. The high F_{BR} can be attributed to the ultra-wide band gap in these materials, where F_{BR} is expected to scale as the 2.5-power of the bandgap[47]. Because of its higher F_{BR} and comparably high v_{sat} for Al-rich AlGaN relative to GaN, UWBG AlGaN is expected to have Johnson figure of merit (JFOM) [17]

20

$$JFOM = \frac{v_{sat}F_{BR}}{2\pi}$$

much higher than GaN. Thus, $Al_xGa_{1-x}N$ -based devices offer many properties that are beneficial for the fabrication of high power, high voltage devices and are expected to play an important role in advancing semiconductor device technology.

For RF devices, two of the main parameters of interest are the unity current gain cutoff frequency, f_T, and unity power gain cutoff frequency, f_{max}. Both these parameters are strong functions of the source and drain resistance, R_s and R_d, which are dependent on the sheet resistance of the channel.

$$f_T = \frac{g_m}{2\pi\left(C_{gs} + C_{gd}\right)\left(1 + \frac{(R_s + R_d)}{R_0}\right) + g_m C_{gd}(R_s + R_d)}$$

$$f_{max} \approx \frac{f_T}{2\sqrt{\frac{R_i + R_s + R_g}{R_0} + (2\pi f_T)R_g C_{gd}}}$$

where g_m, C_{gs}, C_{gd}, R_0, R_i, R_s, R_d, R_g are device transconductance, gate to source capacitance, output resistance, input resistance, source resistance, drain resistance and gate resistance, respectively.

The total resistance is an important parameter for power devices as well, as the Baliga Figure of Merit (BFOM) is inversely proportional to on-resistance, as can be seen below

$$BFOM = \epsilon \cdot \mu \cdot F_{BR}^3 = \frac{4 \cdot V_{BR}^2}{R_{on}}$$

where V_{BR} is the breakdown voltage of the device R_{on} is the on-resistance. ε, μ and F_{BR} represent the dielectric constant, carrier mobility and material breakdown electric field, respectively. Hence, scaling is also critical for fabricating power devices with $Al_xGa_{1-x}N$ channel transistors. Another benefit of scaling is the diminishing size of the integrated circuit (IC) chips, which results in more compact products.

Reports of multiple methods, such as direct metal-semiconductor contact and heterostructure based contacts, exist in the literature for making ohmic contacts to high Al-composition $Al_xGa_{1-x}N$ channel devices, but none have been able to provide a comprehensive solution to this problem. This chapter starts with a review of the traditional metal-semiconductor based contact formation solution that have been explored previously and discusses the limitations of such techniques for high Al-composition AlGaN. The chapter then discusses the details of the novel heterostructure engineered contact formation scheme which forms the essential backbone of forming low resistance contacts to AlGaN channel devices in this book.

2.2. METAL SEMICONDUCTOR BASED APPROACHES

The ohmic contact is a non-rectifying electrical junction that ideally passes the required current without dropping any voltage across it. This is typically achieved in a transistor by forming a metal-semiconductor contact to the source and drain regions with the metal work-function being ideally as close as possible to the semiconductor electron affinity for n-type contacts. In most cases, however, there will be a positive metal-semiconductor energy barrier at the metal-semiconductor interface. With a high enough doping in the semiconductor (typically greater than 10^{19} cm^{-3}), however, the width of the depletion region at the metal-semiconductor interface can be made very thin, on the order of a few nanometers, such that carriers can readily tunnel across such barrier. The heavy doping can be achieved either by annealing the contact, thus causing a reaction between the metal and the semiconductor, and/or heavily doping the semiconductor contact regions.

2.2.1. CONTACT FORMATION TO GaN

The most common ohmic contact formation strategy to n-GaN has been to use metals such as Ti and/or Al with a work function close to the electron affinity of GaN. These metals form low Schottky barriers to n-GaN. Such contact schemes have primarily employed a metallization scheme of Ti/Al/(Ni, Ti, Mo, Pt, or Nb)/Au[48-54]. The two main mechanisms for this low resistance Ohmic contact formation for AlGaN/GaN high electron mobility transistors (HEMTs) have been postulated to be either low Schottky barrier contacts and/or tunneling contacts[55]. High temperature rapid thermal annealing (RTA), typically at a temperature above 800° C, has been used extensively, with great success for

the formation of such contacts. During this process, interfacial reaction between the metals

and the semiconductor results in spike-like structure formation driven by the dislocations

in GaN, which directly touches the two-dimensional electron gas (2DEG)[56]. Besides this,

during the annealing process Ti reacts with GaN, leading to the formation of TiN, which

in turn produces N vacancies (V_N) that act as donors and heavily dopes the metal-

semiconductor interface[56]. This results in ultra-thin depletion regions which are virtually

transparent to electron conduction[57]. The low Schottky barrier coupled with heavy

doping and additional interface doping by annealing, have been successful in

demonstrating contact resistance as low as 0.1 Ω.mm[58].

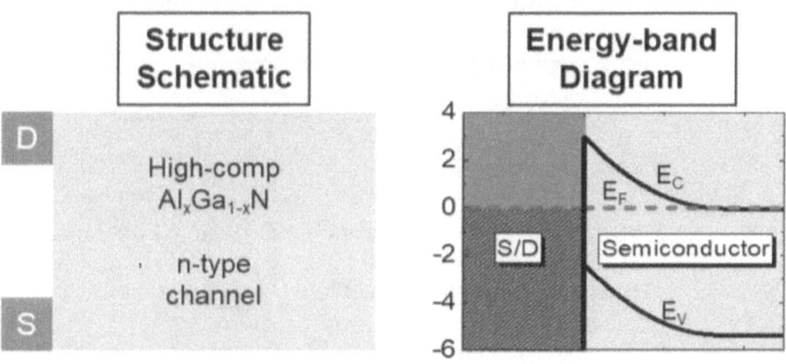

Figure 2.1 (left) Schematic diagram of a metal semiconductor contact to n-type high Al-

composition Al$_x$Ga$_{1-x}$N channel transistor and (right) energy-band diagram under the

source/drain access region of the metal-semiconductor contact layer simulated via 1D

Schrodinger-Poisson simulator

While this approach has worked well with GaN technology, the same success has not been reproducible with high Al-composition $Al_xGa_{1-x}N$ [59]. The primary reason for this is the low electron affinity in these set of materials. Due to significantly lower electron affinity, especially for Al-composition > 0.5, the barrier heights of the metal semiconductor contact are higher for $Al_xGa_{1-x}N$ channel devices than for GaN devices, which results in the formation of rectifying contacts instead of Ohmic contacts[60]. Moreover, this problem is exacerbated with an increase of the Al-content in the channel[35]. This has essentially been the main impediment to the wide adoption of $Al_xGa_{1-x}N$ channel devices that otherwise provide excellent material properties well suited for high performance power and high-power RF transistors. The schematic of a metal-semiconductor contact to a typical high Al-composition $Al_xGa_{1-x}N$ film, together with the energy-band diagram under the source/drain access region, has been shown in Figure 2.1.

2.2.2. Metal-semiconductor based Contact Formation Approaches to High Al-composition AlGaN

Approaches based on metal-semiconductor contact formation to $Al_xGa_{1-x}N$ films have tried to improve the contact resistance by utilizing various metallization schemes such as Ti/Al/Ni/Au, Zr/Al/Mo/Au, V/Al/V/Au, Nb/Ti/Al/Mo/Au etc.[48-54]. Although there has not been any explicit investigation to understand the physical mechanism that could aid the formation of low resistance contacts with these metallization schemes, the aim of these studies was to either explore ways to reduce the metal-semiconductor barrier heights and/or more effectively dope the metal-semiconductor interface. Efforts have also focused on actively introducing [Si^+] as an n-type dopant to reduce the depletion region thickness and increase the tunneling probability by approaches such as Si ion implantation[49, 61], and introduction of SiN_x interlayer between the metal and semiconductor [62]. Such efforts have proven to be useful for cases with $x < 0.5$, where contact resistance as low as 1.65 Ω.mm has been achieved for AlGaN channel metal-oxide semiconductor high mobility field effect transistors (MOSHFET) for $x = 0.40$[36]. However, for Al-composition above 0.5, these methods have not been as successful, with the best reported result being 24.6 Ω.mm.

2.3. REVERSE AL-COMPOSITION GRADED CONTACT LAYER

The approach that is mostly used in this book for contact formation to high Al-composition AlGaN, involves the use of a heavily doped, reverse Al-composition graded intermediate contact layer between the metal and the channel. For this scheme, the AlGaN composition in the contact layer is reduced from the Al-composition in the channel until it becomes GaN at the surface, on which the contact metal is deposited[33]. A schematic of this is shown in Figure 2.2. Since metal contact is being established to GaN, the metal-semiconductor contact resistance is low in these cases[33, 63, 64]. It is of paramount importance, however, to heavily dope the contact layer with n-type dopants such as [Si^+]. This serves two purposes: 1) keep the contact region highly conductive to minimize any additional resistance being introduced by the contact layer and 2) compensate for any negative polarization charge introduced because of the reverse Al-composition grading of the AlGaN in the contact region[64]. Due to the strong polarization effect displayed by (Al)GaN material system, the reverse Al-composition grading produces the negative polarization charge for the typical metal-face growth orientation[65].

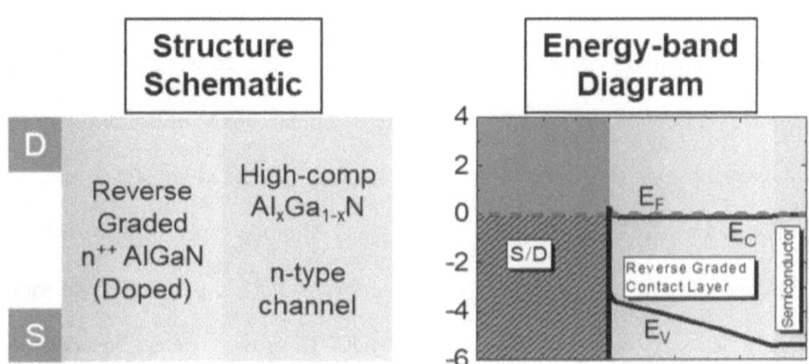

Figure 2.2 (left) Schematic diagram of a reverse Al-composition graded contact layer for high Al-composition AlGaN channel transistor and (right) energy-band diagram under the source/drain access region of the metal-semiconductor contact layer simulated via 1D Schrodinger-Poisson simulator

The first report employing such reverse graded contacts was published in 2015, where the contact layer was implemented for a metal-insulator semiconductor field effect transistor (MISFET)[33]. In this report, molecular beam epitaxy (MBE) was used to grow the transistor channel and the graded contact on a single growth. However, even though contact resistances as low as 0.3 Ω.mm were achieved, the carrier mobility in the channel was low and limited the current density to 60 mA/mm[33]. There are currently no reports of high mobility AlGaN channel transistors grown by MBE reported in the literature.

In contrast, metalorganic chemical vapor deposition (MOCVD) based growth of high Al-composition have routinely achieved high electron mobility with some reports claiming mobilities exceeding 200 cm2/V·s[29, 44, 66]. However, uniform heavy n-type doping via [Si$^+$] of the Al-composition graded contact layers, one of the key requirements of growing low resistance reverse Al-composition graded contact layers, has proven difficult for MOCVD grown contact layers[64]. Multiple experiments performed to gauge the feasibility of such all-MOCVD grown devices have shown this to be true. Previous experiments to establish such in-situ MOCVD grown contact layers have yielded resistances above 10^{-4} Ω.cm2. The likely reason for this difference is that in AlGaN films grown via MOCVD, dopant incorporation is a strong function of the Al-composition of the film, and high dopant incorporation becomes more challenging for higher Al-content AlGaN. As a result, the growth of MOCVD grown reverse-graded contact layers is challenging. In this book, for the first time, we carefully considered the design of the contact layers and have been able to bring down the contact resistance an order of magnitude lower as the following section will illustrate.

Another approach also explored in this chapter is MBE regrown reverse Al-composition graded contact layer on an MOCVD grown channel. Such a hybrid design aims to utilize the high mobility MOCVD grown channel while simultaneously using lower contact resistance achievable via MBE regrown contacts.

2.4. In-Situ MOCVD Grown Reverse Al-composition Graded Contact Layer on MOCVD Grown Channel

Spontaneous and piezoelectric polarization are a unique property of the (Al)GaN heterostructures. The presence of such large polarization charge densities at interfaces, as well as in graded regions is due to the large piezoelectric polarization in these materials originating from the large ionicity of the (Al)Ga–N bonds and the strong spontaneous polarization resulting from the uniaxial nature of the crystal coupled with the non-ideal c/a ratio of the wurtzite structure[67, 68]. When such polar materials are compositionally graded, bound polarization "bulk" charge density are produced. The polarization charge density is given by $D_\pi = -\nabla \cdot P$, where P is the sum of spontaneous and piezoelectric polarization in AlGaN alloys [67]. Typically, as the Al-composition in these materials is compositionally graded from higher Al-composition to lower Al-composition, as is done for a reverse Al-composition graded contact layer, there are bound negative polarization charges formed. This results in the positive curvature in the energy band diagram [33]. Thus to compensate for the negative polarization charge, it is necessary to introduce heavy n-type [Si$^+$] doping. Besides that, additional [Si$^+$] dopants must also be incorporated in the contact layer film so as to lower the contact layer resistance and also to create a flat conduction band profile which can allow unimpeded electron flow[33]. The resistivity of the compositionally graded contact layer, ρ_{graded}, can be described by

$$\rho_{graded} = \int_0^{t_{graded}} \frac{dz}{q \cdot \mu \cdot (N_D(z) - \nabla_z P[x(z)])}$$

where q, μ, $N_D(z)$ and $P[x(z)]$ are the elementary electronic charge, carrier mobility, impurity doping density in the film and the polarization in the graded layer, respectively.

While MBE grown high Al-content AlGaN films consistently achieve high doping density above mid-10^{19} cm^{-3} for x < 0.8, this is not the case for MOCVD grown high Al-composition $Al_xGa_{1-x}N$ films[69-74]. For MOCVD grown $Al_xGa_{1-x}N$ films, dopant incorporation is a strong function of the Al-composition of the film being grown and dopant incorporation becomes more challenging for higher Al-content $Al_xGa_{1-x}N$ – which compounds the difficulty of growing such reverse-graded contact layers[72, 73]. This is different from MBE-grown $Al_xGa_{1-x}N$ films where dopant incorporation is independent of the Al-composition of the film [69, 70]. Thus, while MBE can achieve degenerately doped Al-composition graded contact layer using the same growth condition, a continuous shift in optimum dopant incorporation condition makes growing such uniformly high doped Al-composition graded $Al_xGa_{1-x}N$ films challenging for MOCVD. This indicates contact layer resistance is sensitive to the dopant incorporation, and the activated dopant density needs to be higher than the negative polarization charge density. This is the underlying cause for the high contact resistance reported in previous MOCVD grown reverse Al-composition graded contact layers reported in the literature[37, 63]. To achieve lower contact resistance in MOCVD grown graded contact layers, it is thus necessary to reduce the negative polarization charge density in the graded contact layers, ρ_{graded}, by reducing the factor, $\nabla_z P[x(z)]$ – gradient of the polarization charge.

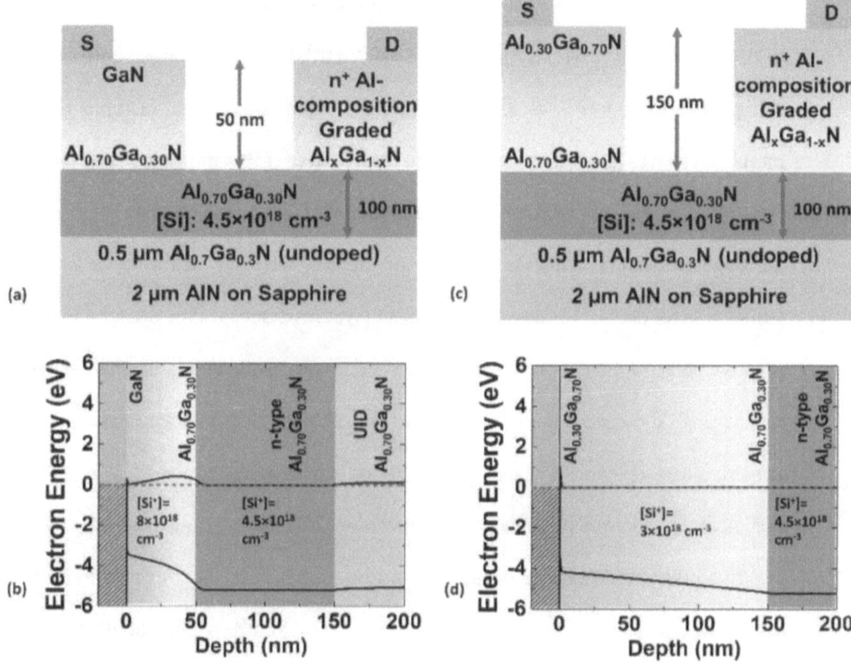

Figure 2.3 (a) Schematic and (b) energy band-diagram of the access region of sample A with 50 nm thick linearly reverse-graded contact layer with x graded from 0.7 to 0 and $[Si^+]$ = 8×10^{18} cm^{-3} and (c) Schematic and (d) energy band-diagram of the access region of sample B with 150 nm thick linearly reverse-graded contact layer with x graded from 0.7 to 0.3 and $[Si^+]$ = 3×10^{18} cm^{-3}

To put this idea into action, two different contact layers were grown on an n-type $Al_{0.70}Ga_{0.30}N$ channel with a doping of 4.5×10^{18} cm^{-3}. One of the contact layers (sample A) was graded from x=0.7 to x=0 over 50 nm, while for another (sample B), the Al-composition was graded from x=0.7 to x=0.3 over 150 nm such that a lower [Si$^+$] concentration can compensate the negative polarization charge density (Figure 2.3). Although a lower x at the top surface of the contact layer will yield a lower metal-semiconductor contact resistance, the contact layer resistance will be higher due to a higher negative polarization charge density due to larger compositional grading and vice versa. A terminating Al-composition of x=0.3 was thus chosen for the second case, since metal-semiconductor contact resistance of low-10^{-5} Ω-cm^2 is expected for $Al_{0.3}Ga_{0.7}N$, which is sufficient for a proof-of-concept demonstration[75]. With a thickness of 150 nm, the negative polarization charge density should be further reduced which will ensure a low resistance contact layer with a uniform $N_D = 3\times10^{18}$ cm^{-3}, which should achievable in MOCVD for x < 0.8[37].

The structures were grown using MOCVD on an AlN on sapphire template with a 100 nm channel with a doping density of 4.5×10^{18} cm^{-3} in the channel and 1×10^{19} cm^{-3} in the graded contact layer. Ti/Al/Ni/Au metal stack was deposited using e-beam evaporation to form source and drain contacts. Contacts on sample B were then annealed at 850 °C for 30 seconds using a rapid thermal annealing (RTA) system to form alloyed contacts, followed by ICP-RIE plasma etch to define device isolation mesas. The active areas of the devices were defined by selectively recessing the graded contact layer between the source and drain contacts. Over-recessing was performed to ensure complete removal of the graded contact layer such that 40 nm of the channel remained after the contact-layer recess. The fabricated samples are shown in Figure 2.3(a-d).

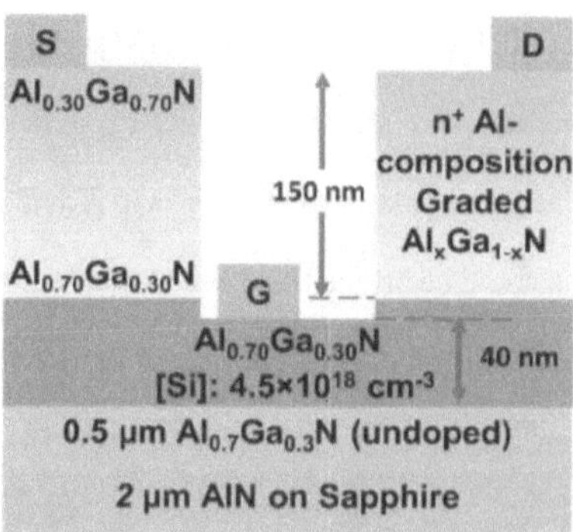

Figure 2.4 Schematic of the fully processed sample B with a 150 nm reverse Al-composition graded contact layer with a channel thickness of 40 nm

Two-terminal current-voltage (IV) measured on the fabricated samples showed that sample A had non-linear IV characteristics – indicative of insufficient compensation of negative polarization charge in the contact layer. On the other hand, sample B displayed linear IV characteristics, which implies that the negative polarization charge in the contact layer has indeed been compensated as expected. Ungated four-terminal Van der Pauw (VDP) structures were used for Hall measurements which showed a channel sheet resistance of 5.6 kΩ/$\square$, sheet carrier density of 1.8×10^{13} cm^{-2} and mobility of 56 cm^2/V·s.

Transfer length measurements (TLM) performed on gated TLM structures with recessed contact layer on sample B yielded a specific contact resistivity of 3.3×10^{-5} $\Omega \cdot cm^2$. TLM and Hall measurements were not performed on sample A, due to the non-linear nature of the contacts. The contact layer resistivity was estimated to be approximately 1×10^{-5} $\Omega \cdot cm^2$ which is higher than MBE grown contact layers [33]. The previous equation for resistivity of graded layer predicts that the incorporated dopant density should be sufficient to compensate for the negative polarization charge and provide low resistance contact. Hence, the higher resistance could be originating from a) non-uniform grading of the contact layer leading to higher localized polarization charge and/or b) lower localized [Si$^+$] incorporation – both of which can lead to high resistance regions. This indicates that further growth optimization is required.

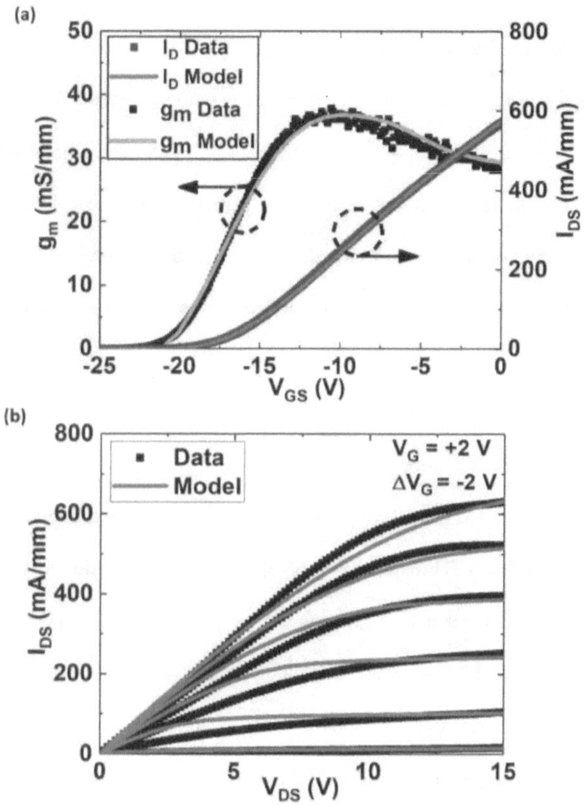

Figure 2.5 (a) Transfer characteristics and (b) family output I–V characteristics for device with gate-length, L_G = 0.6 µm and source-drain spacing, L_{SD} = 1.5 µm. Square symbols represent measured electrical characteristics while solid lines represent simulated characteristics.

Gate metal stacks of Ni/Au/Ni were deposited on sample B using e-beam evaporation to form metal-semiconductor field effect transistor (MESFET) structures shown in Figure 2.4. Transfer IV characteristics (Figure 2.5(a)) measured for devices with L_G = 0.6 µm and L_{SD} = 1.5 µm (V_{DS} = +20 V) showed a pinch-off voltage of -16 V and a maximum transconductance of 38 mS/mm. Output electrical characteristics (Figure 2.5(b)) measured on the same device showed a maximum current density of 635 mA/mm (V_{GS} = +2 V). 2-D technology computer aided design (TCAD) simulator, Silvaco Atlas, was used to model the described device structure[14]. As can be seen, good match was obtained for transfer characteristics and saturated current density. However, some deviation was observed in the slope of the output curves in the linear region – most likely originating from a mismatch between the field dependent mobility model in Atlas and the actual device characteristics. Three-terminal breakdown characteristics of the MESFETs measured at V_{GS} = −20 V for devices with gate-drain spacing, L_{GD} = 0.77 µm, showed no breakdown up to V_{GD} = +220 V, which translates to an average field of 2.86 MV/cm, almost 3× higher than that exhibited by lateral GaN channel devices with similar dimensions. The breakdown is mainly limited by the gate leakage current that is the primary contributor to the drain current in the three terminal breakdown measurement. Thus, the breakdown characteristics can be further improved by the addition of a gate dielectric such as Al_2O_3[63].

2.5. MBE REGROWN REVERSE AL-COMPOSITION GRADED CONTACT TO HIGH AL-COMPOSITION AlGaN CHANNEL TRANSISTORS

The previous work on MOCVD grown reverse Al-composition graded contact layer showed an order of magnitude higher contact resistance compared to the MBE grown contact layer. To explore the potential integration of highly conductive MBE regrown contact layers on high mobility MOCVD grown channel, MBE regrowth was performed on MOCVD grown n-doped films. Two separate approaches are feasible: (a) top regrown MBE contacts on MOCVD grown channels and (b) side regrown MBE contacts in MOCVD grown channels. Both these approaches are schematically shown in Figure 2.6. Although top regrowth process is a considerably simpler approach than side regrowth process, side regrowth was primarily focused on in this work since (a) this method is suitable for films with different spatial electron distribution including two-dimensional electron gas (2DEG) in HEMTs; (b) no polarization charge at the regrowth interface since the primary electron injection occurs in the m-plane which will not have polarization charge due to composition mismatch between MOCVD grown channel and MBE regrown contact layer; and (c) composition matching is relatively easier since there is a larger contact area between the channel and the contact layer.

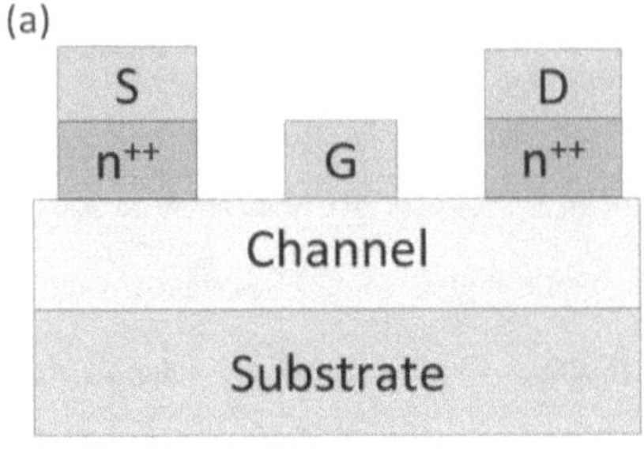

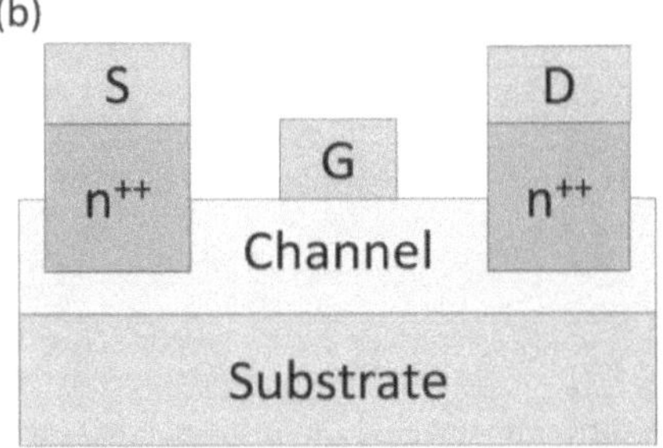

Figure 2.6 (a) Top MBE contact regrowth scheme on MOCVD grown channel and (b) side MBE contact regrowth scheme on MOCVD grown channel.

2.5.1. MBE Top Regrown Contact to High Al-Composition AlGaN channel POLFETs

For this study, metal organic chemical vapor deposition (MOCVD) was used to grow an upward composition-graded high Al-composition AlGaN channel, while a reverse composition-graded n^{++} AlGaN layer was regrown via molecular beam epitaxy (MBE) on the MOCVD grown channel to minimize abrupt conduction band offsets and facilitate low resistance non-alloyed ohmic contact formation. High mobility in MOCVD-grown channels and the ability to achieve heavy n-type doping toward low contact resistance in MBE-grown graded contact layers were the motivating factors behind using this MOCVD-MBE hybrid growth.

For MBE contact optimization, three samples were prepared with the same MOCVD grown channel layer on sapphire templates. The stack consisted of an AlN nucleation layer followed by pseudomorphic growth of 500 nm thick $Al_{0.65}Ga_{0.35}N$ buffer layer, and 60 nm thick AlGaN channel with Al-composition grading from ~65% to ~88% ($[Si^+] = 1 \times 10^{18}$ cm^{-3}). 50 nm reverse Al-composition graded contact layers were grown using MBE with a high Si-doping concentration of 1×10^{20} cm^{-3}. The starting composition of the MBE regrown graded layer was varied, and the ending composition was kept constant at 0% (or GaN). The three samples, S84, S70, and S60, had linearly graded contact layers with Aluminum mole-fraction of 84%, 70%, 60%, respectively at the bottom, and 0% at the top.

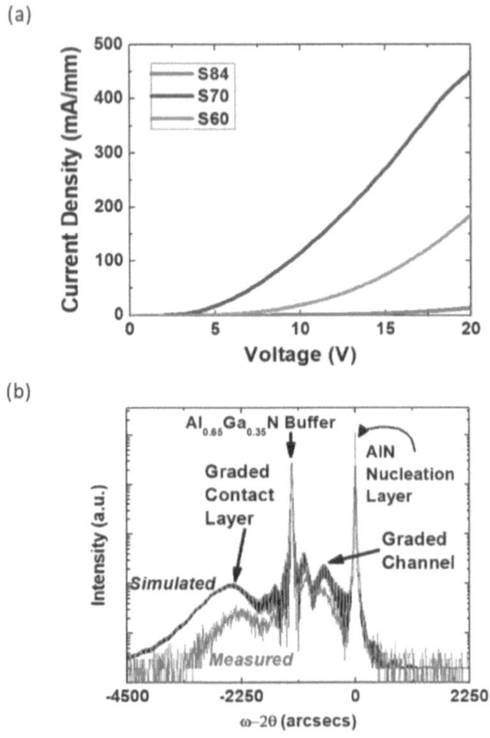

Figure 2.7 (a) Two-terminal IV characteristics of S84, S70 and S60 (b) HRXRD on S70 to confirm the composition of the graded channel and contact layer

E-beam evaporation was used to deposit a non-alloyed metal stack of Ti/Al/Ni/Au for source/drain contacts, followed by 160 nm mesa structure formation via Cl$_2$-based ICP-RIE etching to electrically isolate the devices. Device access and intrinsic regions were defined by removing the 50 nm thick MBE-regrown contact layer using a calibrated low-power (5W) Cl$_2$-based plasma etch. Incomplete recessing of the n^{++}-contact layer can degrade Schottky gate characteristics, hence a 15 nm over-etch was performed. This defined a channel thickness of 45-nm with an Al-composition varying from 65% to 82%. This also improved the aspect ratio of the device thus reducing short channel effects.

Two-terminal measurements across ungated transistor structures were performed to estimate the optimal contact (shown in Figure 2.7(a)). The low current density observed in S84 (MBE and MOCVD growth Al-composition are closely matched at the interface) is due to the formation of higher Al-composition AlGaN channel interlayer which introduces a resistive region in the contact and/or channel layer. The presence of the higher Al-composition layer was confirmed using high resolution transmission electron microscopy (HRTEM) as shown in Figure 2.8. The non-ohmic IV characteristics of S70 and S60 results from the conduction band offset (ΔE_C), due to composition difference between the MBE contact layer and MOCVD channel layer due to the formation of the higher Al-composition AlGaN interlayer.

As can be seen from Figure 2.7(a), S70 was found to have the highest two-terminal current density among all these three samples. Subsequently discussion on only this sample is provided in the rest of this section. High-resolution XRD diffraction pattern was measured on the sample with a BEDE D1 High Resolution Triple Axis X-ray Diffraction System (Figure 2.7(b)). The measured XRD pattern matches the simulated pattern (including contact layer). The equilibrium energy-band diagrams were calculated under the device access regions (Figure 2.9(a)) and the recessed gate region (Figure 2.9(b)). The band profiles have been simulated using a Schrodinger-Poisson solver[30]. The Al-composition grading in the heavily doped n⁺⁺-contact region enables a flat conduction band profile between the metal contact and the high composition AlGaN. Gate metal stack of Ni/Au/Ni was thus deposited on S70 via e-beam evaporation. Schematic of the fully processed device structure is shown in Figure 2.10(a) while Figure 2.10(b) shows the top-view scanning electron microscope (SEM) image of the device.

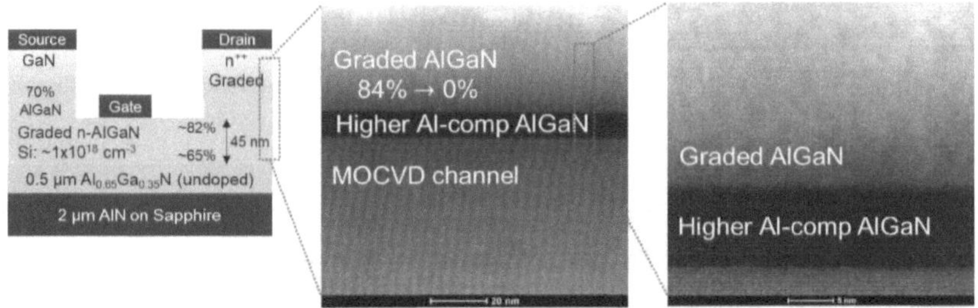

Figure 2.8 HRTEM imaging showed that there was a higher Al-composition AlGaN interlayer present in the S84.

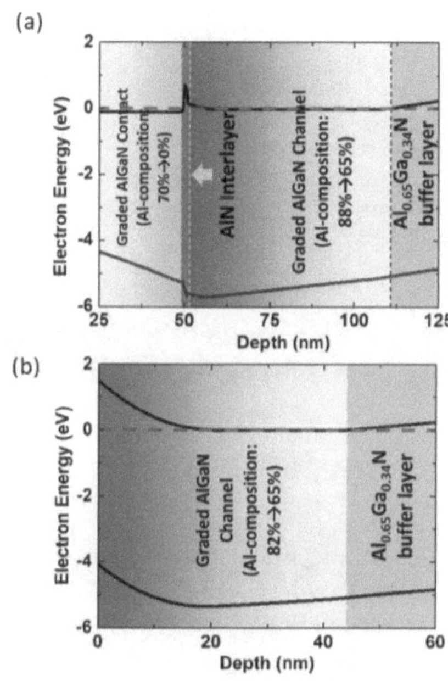

Figure 2.9 Calculated energy-band diagram for S70 (a) under the device access region and (b) recessed gate region

(a)

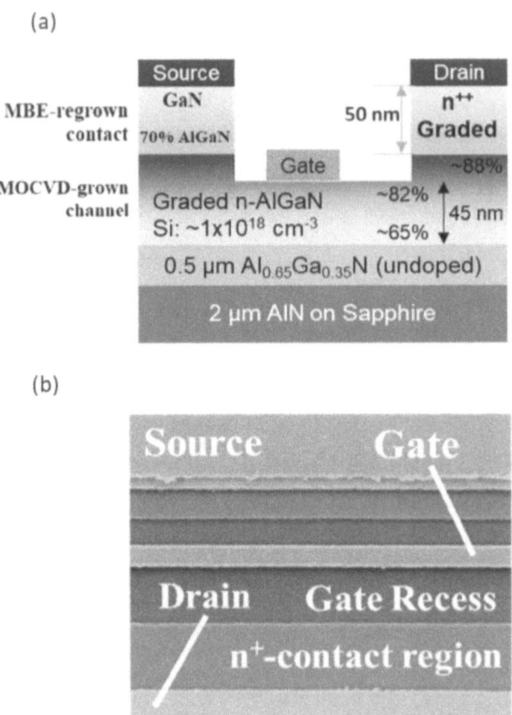

(b)

Figure 2.10 (a) Schematic cross-section and (b) top-view SEM image of fully processed sample S70

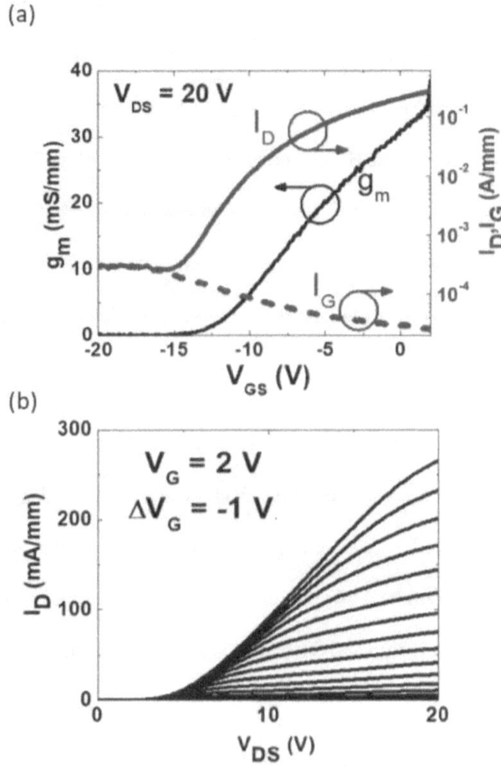

Figure 2.11 (a) Transfer and (b) output characteristics for a PolFET device with L$_G$ = 0.8 μm, L$_{GD}$ = 2.0 μm and L$_{GS}$ = 0.9 μm

46

Electrical characterization was performed on the devices using an Agilent B1500 parameter analyzer (Figs. 2.11 and 2.12). Figure 2.11(a) shows the transfer IV characteristics of a PolFET device with L_G = 0.8 µm, L_{GD} = 2.0 µm and L_{GS} = 0.9 µm (dimensions measured using SEM), measured at V_{DS} = 20 V. The pinch-off voltage is around -12 V with an I_{ON}/I_{OFF} ratio of ~10^3 with gate leakage being the main contributor to the I_{OFF} current. Surface damage due to gate recess etch is the most likely cause of the high gate leakage current. This limited the three-terminal breakdown voltage to 35 V.

As mentioned before, the non-ohmic IV characteristic at the triode region of the I_D-V_D plot is a result of the composition difference between the MBE contact layer and MOCVD channel layer at the MBE-MOCVD interface. At low V_{DS}, the conduction band offset, ΔE_C, introduced by the composition mismatch, impedes current flow which manifests as a voltage offset. However, for V_{DS} > ~4V, electrons tunnel through the barrier as it becomes sufficiently thin and the contacts behave like ohmic contacts with an effective contact resistance of ~7 Ω.mm (estimated from the $I_D - V_D$ curves in Figure2.11(a)). The output electrical characteristics of the devices are shown on Figure 2.11(b). A maximum current density, $I_{DS,max}$, of 265 mA/mm is observed (V_{GS} = +2 V). This is lower than the maximum two-terminal current because of the depletion under the gate introduced by the gate metal stack. The output characteristics show good modulation of the channel with pinch-off around V_{GS} = -12 V, as expected.

Capacitance-voltage profile (Figure 2.12(a)) obtained from CV measurement performed on 100µm × 100µm gated TLM (GTLM) structures show an extracted doping density (Figure 2.12(b)) of ~3.5×10^{18} cm^{-3} (integrated charge density $\approx 1.0 \times 10^{13}$ cm^{-2}), with a sharp spike in carrier concentration near the bottom of the channel, probably

originating from an unintentional abrupt Al-composition change at the start of the channel

growth, leading to a large polarization sheet charge. Hall measurement on Van der Pauw

(VDP) structures gave a total integrated charge density, n_{Hall}, of 1.1×10^{13} cm^{-2} (consistent

with charge density obtained from CV profile), and Hall mobility, μ_{Hall}, of 85 cm^2/V-s.

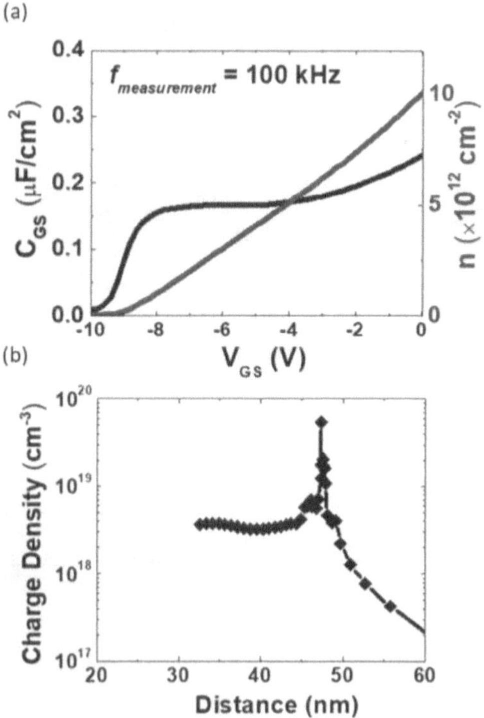

Figure 2.12 (a) Capacitance-voltage profile and the integrated charge density (b) Doping

density extracted from the capacitance-voltage profile.

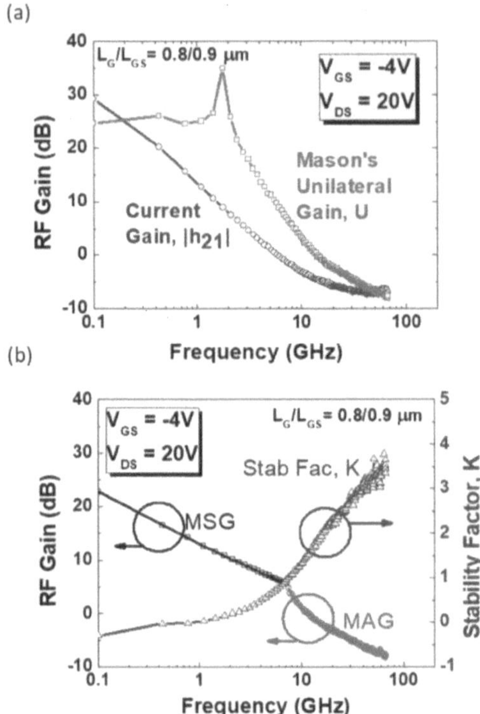

Figure 2.13 (a) Extrinsic RF current and power gain characteristics and (b) maximum stable gain (MSG) / maximum available gain (MAG) and stability factor, K, at $V_{GS} = -4$ V, $V_{DS} = 20$ V

The small signal RF performance was evaluated on the same transistor with an E8361A PNA Network Analyzer. On-wafer calibration was carried out using short-open-load-through (SOLT) off-wafer impedance standards in the frequency range from 100 MHz to 67 GHz. The S-parameters measured with a drain-source voltage, V_{DS} = +20 V and a gate-source voltage, V_{GS} = -4 V show a peak extrinsic f_T of 5.4 GHz and a peak extrinsic f_{MAX} of 14.2 GHz (Figure 2.13(a)). The maximum stable gain (MSG) / maximum available gain (MAG) and the stability factor have been plotted in Figure 2.13(b) which shows a stability factor cross-over point (K = 1) of 7.33 GHz.

2.5.2. STUDY OF MBE REGROWN REVERSE AL-COMPOSITION GRADED CONTACT TO HIGH AL-COMPOSITION ALGAN CHANNEL TRANSISTORS

Side regrowth experiments were performed on MOCVD grown n-type ($[Si^+]$ = 5×10^{18} cm^{-3}) $Al_{0.60}Ga_{0.40}N$ films. The selective area regrowth process flow of the fabrication of MBE side-regrown contacts on the MOCVD grown films is as follows. 500 nm of silicon oxide (SiO_2) was first deposited on the MOCVD-grown channel layer which was then patterned and selectively removed. The SiO_2 was then used as hard mask to form 50 nm deep pits on the MOCVD grown channel. Heavily $[Si^+]$-doped (1×10^{20} cm^{-3}) $Al_{0.60}Ga_{0.40}N$ (50 nm) was then blanket regrown in a molecular beam epitaxy (MBE) chamber which was capped with a heavily $[Si^+]$-doped reverse Al-composition graded AlGaN layer of thickness of 50 nm and a doping of 1×10^{20} cm^{-3}. The purpose of the reverse composition-graded n^{++} AlGaN layer is to minimize abrupt conduction band offsets and facilitate low resistance non-alloyed ohmic contact formation. Crystalline films were only formed in the pits while amorphous AlGaN was formed on SiO_2. Following MBE contact regrowth, SiO_2 regrowth mask was removed using a diluted buffered oxide etch (BOE) solution, and a metal stack of Ti/Al/Ni/Au [20/120/30/100 nm] was deposited on the regrown contact regions on the pits via an e-beam evaporator to form ohmic contacts. The process up to the source/drain metal deposition is shown in Figure 2.14.

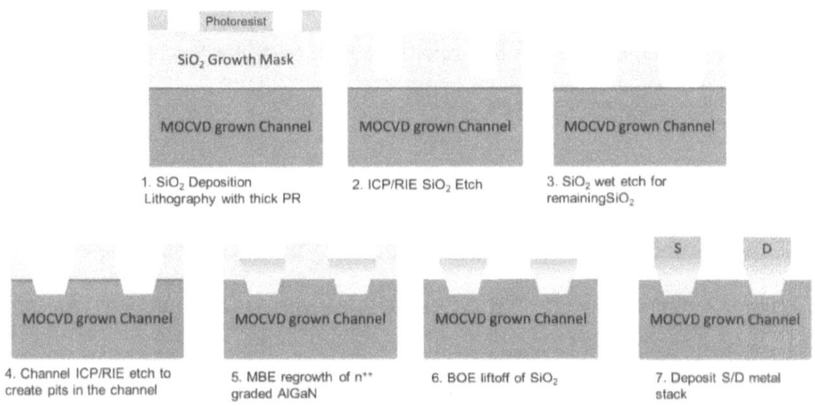

Figure 2.14 Process flow of side regrown reverse Al-composition graded contact layer for applications in the fabrication of MOCVD grown AlGaN channel devices

Following the S/D metal deposition, device isolation was performed by inductively coupled plasma – reactive ion etching (ICP-RIE) plasma etching system with an RIE power of 30 W and a pressure of 5 mTorr. The final device structure is shown in Figure 2.15(a). High-resolution x-ray diffraction (XRD) spectra (BEDE D1 High Resolution Triple Axis X-ray Diffraction System) was used to confirm close composition match between the MBE regrown contact layer (XRD was performed on a co-loaded AlN on sapphire template) and MOCVD grown channel (Figure 2.15(b)). Electrical characterization of the devices was performed using an Agilent B1500 parameter analyzer. Two-terminal IV measurements performed on contact regrown samples showed a non-linear IV as shown in Figure 2.15(c) below:

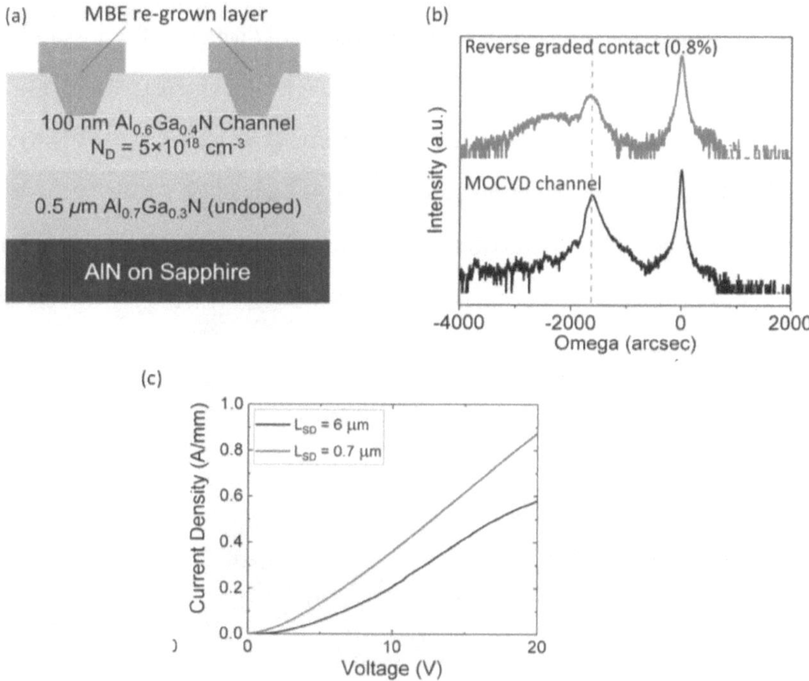

Figure 2.15 (a) Schematic of the MBE side regrown n-doped MOCVD film, (b) XRD on MOCVD grown channel and reverse Al-composition graded contact layer grown on an AlN on sapphire substrate and (c) two terminal IV characteristics of the MOCVD grown channel with MBE regrown contacts.

Due to the close composition match between the MBE regrown contact layer and the MOCVD grown channel, the IV curves were expected to be linear. However, TEM imaging performed on the regrown sample again showed the presence of an interlayer (Figure 2.16). The darker contrast represents higher Al-composition of the film. As can be seen, there is a 1-2 nm thick interlayer of what appears to be higher composition AlGaN. To understand the source of non-linearity, transmission electron microscopy (TEM) was performed on a sample from the same wafer with 12 nm of MBE regrown $Al_{0.60}Ga_{0.40}N$ contact layer regrown with conditions same as the previous sample (Figure 2.17). The same experiment was also performed on the same substrate but this time GaN was regrown instead of AlGaN. Interestingly, the higher Al-composition AlGaN interlayer was not present in the case of the GaN growth, which indicates that interlayer is forming because of the growth of AlGaN itself and not because of any oxide layer that might present in the growth substrate.

Figure 2.16 Schematic and the TEM results from MBE side regrown contact layers. 1-2 nm of interlayer was present in the regrown samples which was likely the origin of the non-linear IV characteristics.

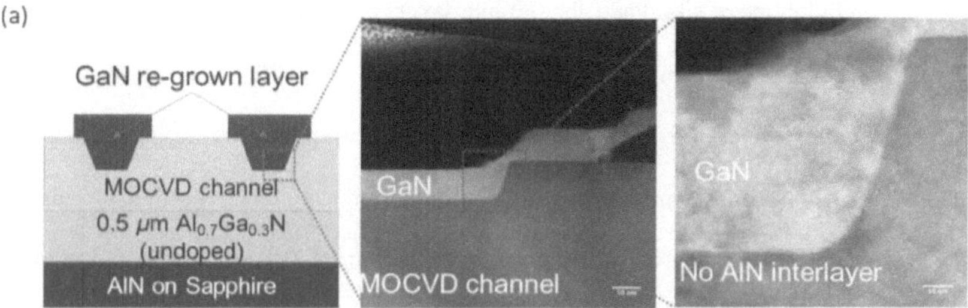

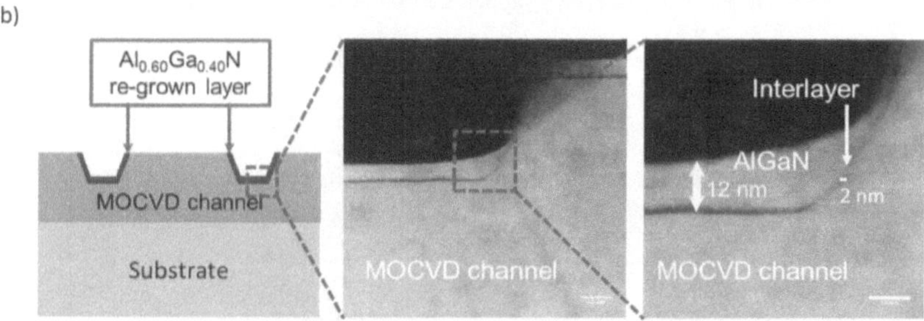

Figure 2.17 Growth experiment to understand the source of the higher Al-compostion AlGaN interlayer. (a) GaN was regrown on an $Al_{0.60}Ga_{0.40}N$ substrate and (b) $Al_{0.60}Ga_{0.40}N$ was regrown on an $Al_{0.60}Ga_{0.40}N$ substrate

To confirm that the interlayer is the source of the non-linear IV, two-terminal IV was simulated in Silvaco Atlas using a device structure same as the experiment above. The interlayer thickness was kept at 1 nm. The simulation was performed for various interlayer composition mismatches, while the contact layer composition was kept the same as the channel layer. The doping the channel was assumed to be 5×10^{18} cm^{-3} while the MBE regrown contact layer was assumed to have a doping of 1×10^{20} cm^{-3}. Two separate doping

54

conditions were simulated for the interlayer, one with a doping concentration of 1×10^{20} cm^{-3} like the regrown contact layer doping and 1×10^{18} cm^{-3} lower than the doping concentration of the regrown contact layer. The results are shown in the Figure 2.18 below. As can be seen, the interlayer has a significant impact on current injection for composition mismatch ($Al_xGa_{1-x}N$) for composition mismatch ≥ 0.2. As expected, a decrease in doping in the interlayer further reduces the current injection. Based on the simulation, the Al-composition of the interlayer appears to be at least 0.80. It also appears that the doping concentration in the interlayer is likely lower than the doping concentration of the MBE regrown contact layer, since the turn-on voltage in the actual device is similar to the turn-on voltage obtained for the simulated results with a lower doped interlayer.

The contact resistance for low bias ($V_{DS} = 0$ V) and high bias conditions ($V_{DS} = 10$ V) were also simulated and shown in Figure 2.19(a) and Figure 2.19(b) respectively. The contact resistance becomes noticeably lower as carrier tunneling processes become dominant at higher drain bias.

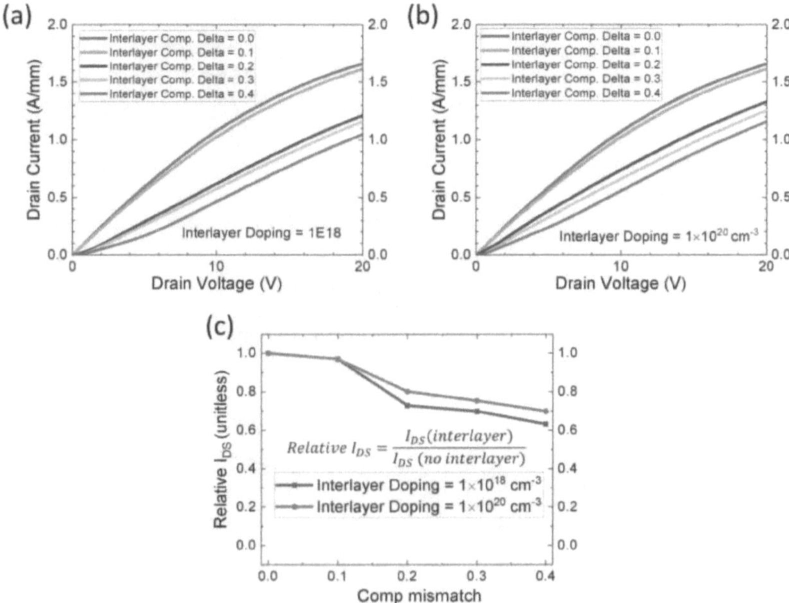

Figure 2.18 Simulated two terminal IV characteristics for MBE regrown contact layer on MOCVD grown channel with interlayer doping concentration of (a) 1×10^{18} cm^{-3} and (b) 1×10^{20} cm^{-3}. (c) Relative two terminal current density as a function of the doping composition mismatch

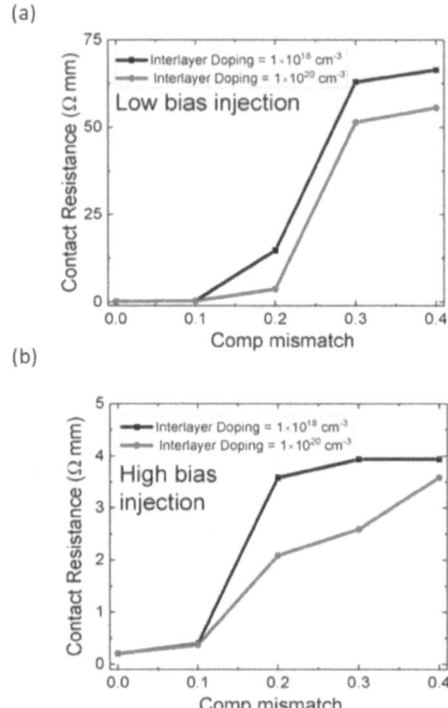

Figure 2.19 Simulated contact resistance for various composition mismatch between interlayer and channel for doping of 1×10^{18} cm^{-3} and 1×10^{20} cm^{-3} under (a) low bias injection ($V_{DS} = 0$ V) and (b) high bias condition ($V_{DS} = 10$ V)

For this study, a 100 nm thick n-type $Al_{0.60}Ga_{0.40}N$ sample with a doping of 2×10^{18} cm^{-3} was used. The active layers were grown via low-pressure (LP) MOCVD on 3 μm thick high quality AlN-(0001) sapphire templates. These template layers were also deposited by LP-MOCVD at growth temperatures close to 1250 ℃. The epilayer consisted of a 500 nm thick undoped i-$Al_{0.60}Ga_{0.40}N$ buffer layer grown pseudomorphically over the entire two-inch diameter AlN template at a growth temperature close to 1100 ℃.

Ohmic contact to the MOCVD-grown n-doped AlGaN layer was achieved via selective area regrowth by MBE, like the process described in the previous MBE regrowth section. After the MBE contact regrowth process, the SiO_2 hard mask was removed and a metal stack of Ti/Al/Ni/Au [20/120/30/100 nm] was deposited on the regrown contact regions via an e-beam evaporator to form ohmic contacts. This was followed by device isolation by inductively coupled plasma – reactive ion etching (ICP-RIE) system with an RIE power of 30 W and a pressure of 5 mTorr.

Device gates were formed by the deposition of Ni/Au [60/100 nm] metal stack via e-beam evaporation. As can be seen (Figure 2.20), for low V_{DS}, the IV characteristics are non-linear, however, as the V_{DS} is increased, the contact resistance improves slightly due to better electron injection in the channel through the interlayer. This is consistent with the simulated results obtained in the previous section.

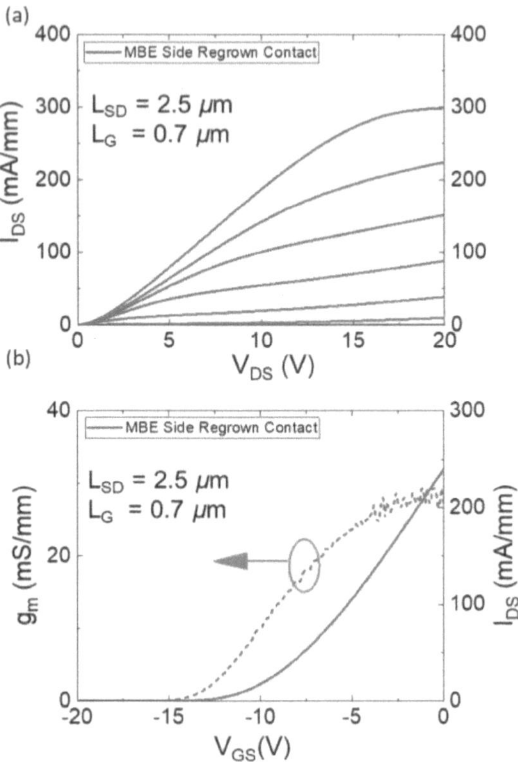

Figure 2.20 (a) Output and (b) transfer curve profile of MBE side regrown contact MESFETs

2.6. SUMMARY

In this chapter, we have developed several state-of-the-art techniques for the formation of low resistance contacts to high Al-composition AlGaN channel transistors. This includes the development of novel in situ slow-graded MOCVD grown contacts to MOCVD grown channels, the development of regrown MBE contacts (side and top) to MOCVD grown channels. Although further improvement in the techniques is still possible, record low contact resistances have been demonstrated in this chapter using the developed techniques.

CHAPTER 3:

LOW FIELD TRANSPORT AND MOBILITY ENGINEERING IN HIGH AL-COMPOSITION ALGAN CHANNEL TRANSISTORS

3.1. BACKGROUND

One of the key challenges of fabricating high-performance RF and power devices using materials such as β-Gallium Oxide (β-Ga2O3) and high Al-composition AlGaN is the low mobility in these materials. In the case of AlGaN, the main reason for the low mobility is the presence of high levels of alloy scattering especially for AlGaN with Al-composition, x, roughly between 0.2 to 0.8. As a result, the theoretically predicted highest mobility achievable in this range of AlGaN is of the order 200 cm2/V.s. Correspondingly, the sheet resistance, RSH, is high in these materials with

$$R_{SH} = \frac{1}{en_s\mu}$$

where e is the elementary electronic charge, n_s is the sheet charge density and μ is the carrier mobility.

For power devices, the mobility determines the on-resistance of the device that directly correlates to the power dissipation in these devices and thus the overall system power efficiency. Mobility is an important parameter for RF devices as well because, for the same bias conditions in devices with similar dimensions, higher mobility devices yield higher transit velocity ergo lower transit delays. Although velocity in low mobility devices can be significantly boosted by physically shrinking the device dimensions with minimal performance degradation, possessing higher mobility will inherently present an advantage. In this chapter, we therefore investigate the low field transport properties in a high Al-composition $Al_{0.70}Ga_{0.30}N/Al_{0.50}Ga_{0.50}N$ HEMT. We also present a potential pathway to significantly increase mobility in AlGaN channel transistors by utilizing multi-channel configuration with reduced charge per channel.

Low-pressure (LP) MOCVD was used to grow the active layers for the reported device structures on 3 μm thick high quality AlN-(0001) sapphire templates. These template layers were also deposited by LPMOCVD at growth temperatures close to 1250 ℃. Their RMS surface roughness was measured by atomic-force microscope (AFM) to be 5.78 nm, as shown in Figure 3.1. The epilayer consisted of a 250 nm thick undoped i-$Al_{0.50}Ga_{0.40}N$ buffer layer grown pseudomorphically over the entire two-inch diameter AlN template at a growth temperature close to 1100 ℃. This was followed by the growth of a 50 nm thick $[Si^+]$-doped n-type $Al_{0.70}Ga_{0.30}N$ layer, with a doping concentration of 2×10^{18} cm^{-3} on top of the buffer layer. A reverse Al-composition graded contact layer was grown on top of $Al_{0.70}Ga_{0.30}N$ layer and was effectively graded from $Al_{0.70}Ga_{0.30}N$ to $Al_{0.30}Ga_{0.70}N$ over 70 nm with heavy $[Si]^+$ doping. The AlGaN contact layer was terminated at $Al_{0.30}Ga_{0.70}N$ because good alloyed ohmic contact can be formed at his composition and also to ensure that the negative polarization change introduced because of the reverse Al-composition grading was such that can be compensated by heavy $[Si]^+$ doping.

Source and drain ohmic contacts were established on the reverse Al-composition graded contact layer by electron-beam evaporation of a metal stack of Ti/Al/Mo/Au, followed by rapid annealing at 875 °C for 45 s. Next, device isolation was performed by selective etch of conductive layers between the devices using Cl_2-based ICP-RIE etching. Active regions were subsequently created by a performing a selective contact layer removal between the source and drain regions by Cl_2-based ICP-RIE etch. A low power etch was carried out to ensure minimal plasma damage to the epitaxial layers. This is necessary to form low-leakage gates on the devices. An over-etch of 30 nm into the barrier layer was also performed to ensure that contact layer was completely removed which retained 20 nm of $Al_{0.7}Ga_{0.3}N$ barrier above the channel region. This ensured that the doped barrier layer was fully depleted which removed potential parallel conduction pathway between source and drain. Gate metal stack consisting of 60 nm of Ni and 100 nm of Au was subsequently deposited on the recessed regions between source and drain to form the gate contact. A schematic structure of the fabricated $Al_{0.7}Ga_{0.3}N/Al_{0.5}Ga_{0.5}N$ HFETs and the equilibrium energy-band diagrams calculated using self-consistent one-dimensional Poisson-Schrodinger solver, under the access regions (source/drain) is shown in Figure 3.2.

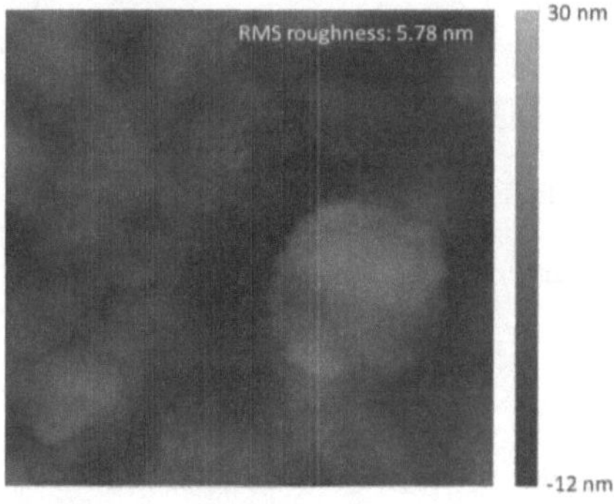

Figure 3.1 AFM scan of the surface of the sample showed an RMS surface roughness of 5.78 nm.

Contact resistance measured using transfer length method (TLM) was found to be 3.9 Ω.mm with a sheet resistance of 7 kΩ/$\square$ which gives a specific contact resistance, ρ_c, of 2.1×10^{-5} Ω.cm^2. Two terminal output characteristics measured on fat field effect transistors (FATFETs) with a source to drain spacing of 18 μm and gate length of 14 μm is shown on Figure 3.3(a). The linear increase in the source to drain current with a corresponding increase in source to drain voltage indicates that the metal-semiconductor contact is ohmic in nature. Hall measurement performed on ungated cloverleaf shaped Van der Pauw (VDP) structures showed a sheet charge density (n_s) of 8.8×10^{12} cm^{-2} and an electron mobility of 120 cm^2/V.s. Sheet charge density obtained from capacitance-voltage (CV) measurements were found to be 5.5×10^{12} cm^{-2} at both room temperature and low-temperature (T = 77 K) (Figure 3.3(b) and (c)). The invariance of the sheet charge density with temperature indicates that the sheet charge exists in a two-dimensional electron gas (2DEG) configuration and hence it is reasonable to conclude that current conduction in the channel is solely through the 2DEG.

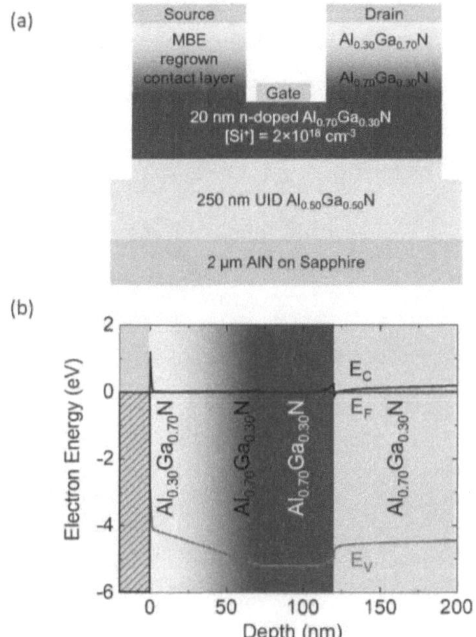

Figure 3.2 (a) Schematic structure of the fabricated MOCVD grown Al$_{0.70}$Ga$_{0.30}$N/Al$_{0.50}$Ga$_{0.50}$N HEMT and (b) under the gate energy band diagram of the fabricated transistor structures.

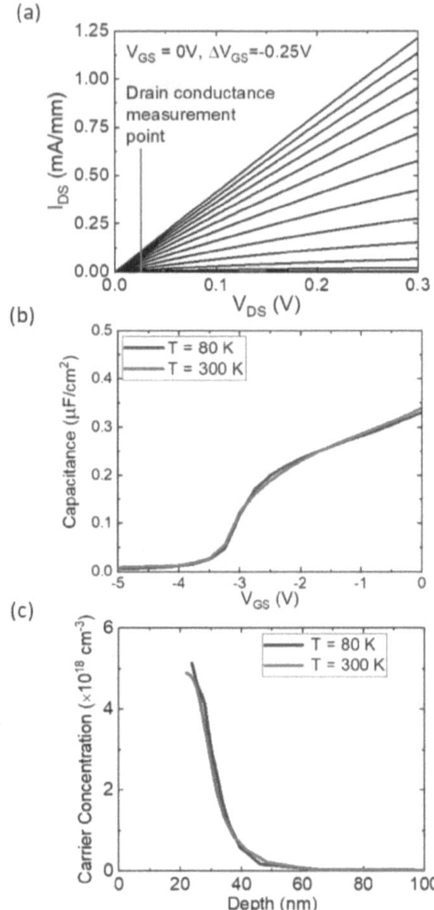

Figure 3.3 (a): Output characteristics at low drain bias condition, (b) room temperature (T = 300 K) and low temperature (T = 77 K) capacitance voltage profile and (c) extracted carrier concentration at both room temperature and low temperature for a 14 μm by 100 μm FATFET structure

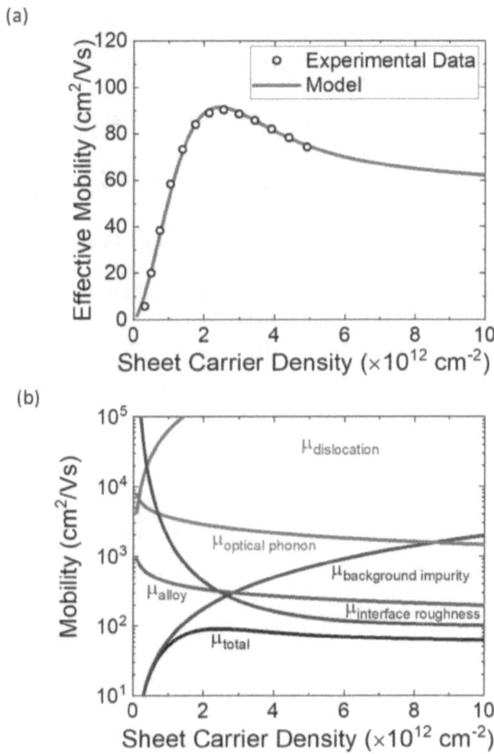

Figure 3.4 (a) Extracted experimental and modeled effective mobility as a function of sheet carrier density and (b) modeled contributions from individual scattering mechanisms.

To estimate the low-field mobility, the effective mobility in these devices has been measured. The relationship between low-field mobility ($\mu_{effective}$) and the device transconductance (g_{ds}), channel charge (Q_n), gate length (L_G) and device width (W) at low drain bias conditions is given by $\mu_{effective} = \frac{L_G \cdot g_{ds}}{W \cdot Q_n}$[60]. Figure 3.4(a) shows the extracted effective mobility as a function of the sheet charge density, as measured on a FATFET with L_{SD} = 18 μm and L_G = 14 μm measured at a drain bias, V_{DS} = 0.025 V. Figure 3.4(b) shows the modeled contributions from the scattering mechanisms that produces the observed mobilities at various sheet charge densities on a log-linear plot. The next section discusses the modeling study performed to produce the shown fit.

3.3. MODELING DETAILS

Analytical models of various scattering mechanisms that included optical phonon scattering, alloy disorder scattering, charged dislocation scattering, background impurity scattering and interface roughness scattering were used to fit experimentally obtained mobility.

The alloy scattering-limited mobility was calculated using a previously derived expression[65].

$$\mu_{ALLOY-DISORDER-SCATTERING} = \frac{e\hbar^3}{(m^*)^2 V_0^2 \Omega_0 (1-x)x} \times \frac{16}{3b}$$

where m^* is the electron effective mass, V_0 is the alloy scattering potential for AlGaN alloys reported earlier to be 1.8 eV, Ω_0 is the unit cell volume calculated using lattice and elastic constants varied linearly between AlN and GaN, x is the Al mole fraction, and $b \sim n_s^{1/3}$ (n_s: 2DEG sheet density) is the variational parameter for Fang-Howard wavefunction, chosen at the minimum energy.

The optical phonon scattering-limited mobility was calculated using[65]

$$\mu_{OPTICAL-PHONON} = \frac{2q_0\hbar^2 F(y)}{e(m^*)^2 \omega_0 N_B(T) G(k_0)}$$

where q_0 is the optical phonon wave-vector, $N_B(T)$ is the Bose-Einstein distribution function, ω_0 is the optical phonon frequency and $G(k_0)$ is the screening form factor.

The charged dislocation scattering is given by[65]:

$$\mu_{DISLOCATION} = \frac{2e\pi\hbar^3 k_F^3}{N_{dis} m_{eff}^2 \int_0^{2k_F} |V(q)|^2 \dfrac{q^2}{\sqrt{1-\left(\frac{q}{2k_F}\right)^2}}}$$

where k_F is the Fermi waveventor, N_{dis} is the areal density of dislocations, m_{eff} is the effective carrier mass, $V(q)$ = Perturbation potential matrix element and, q is the scattering wavevector.

The ionized impurity scattering is provided by [65]

$$\mu_{IONIZED-IMPURITY} \approx \frac{4(2\pi)^{5/2}\hbar^3(\epsilon_0\epsilon_r)^2}{(m^*)^2 e^3} \times \frac{n_s^{3/2}}{N_{imp}}$$

where N_{imp} is the concentration of ionized impurities and n_s is sheet charge density.

Finally, the interface roughness scattering is provided by [65]

$$\mu_{INTERFACE-ROUGHNESS} = \frac{2(\epsilon_0\epsilon_r)^2\hbar^3}{\Delta^2 L^2 e^3 (m^*)^2 \left(\frac{1}{2}n_s\right)^2} \times \frac{1}{\int_0^1 du \dfrac{u^4 \exp(-k_F^2 L^2 u^2)}{\left(u+G(u)\frac{e\tau_F}{2k_F}\right)^2 \sqrt{1-u^2}}}$$

$$where\ u = \frac{e}{2k_F}$$

Mobility as a function of sheet charge density obtained from each of the individual scattering mechanisms are shown in Figure 3.4(b) on a log-linear plot. The dominant scattering mechanism was found to be due to background impurity scattering (impurity concentration: 1.45×10^{18} cm^{-3}) for low sheet charge density while at higher sheet charge densities, interface roughness limits the net mobility in the 2DEG (rms roughness: ~5.95 nm). Remarkably, alloy scattering was not the limiting factor in the samples investigated. Both background impurity scattering, and interface roughness scattering is expected to improve as AlGaN MOCVD growth technology improves and matures. However, ultimately, with improved material quality, alloy disorder scattering will be the limiting factor as this is an inherent scattering mechanism present in ternary alloys such as AlGaN.

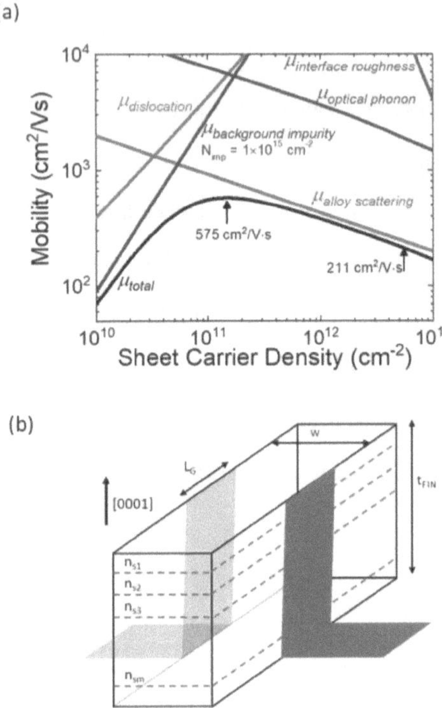

Figure 3.5 (a): Simulated mobility as a function of sheet carrier density for an optimally grown Al$_{0.70}$Ga$_{0.30}$N/Al$_{0.50}$Ga$_{0.50}$N HEMT and (b) Schematic of a double-gated multi-channel AlGaN channel HEMT with low sheet channel charge density per channel to increase 2DEG mobility

For a 2DEG in an optimally grown $Al_{0.7}Ga_{0.3}N/Al_{0.5}Ga_{0.5}N$ HEMT (charged dislocation density: $1 \times 10^8 cm^{-2}$, background impurity concentration: $1 \times 10^{15} cm^{-3}$ and interface roughness: 1 nm), the mobility is expected to exceed 200 cm^2/V.s for a sheet charge density of 5.5×10^{12} cm^{-2} (Figure 3.5(a)). It is interesting to note, that the alloy disorder scattering is proportional to the one-third power of the sheet carrier density in the channel ($\mu_{alloy} \propto \frac{1}{n_s^{1/3}}$)[76]. The inverse relationship between alloy scattering mobility and sheet carrier density is expected since carriers are more sensitive to alloy fluctuations at higher concentrations[76]. As a result, at lower carrier densities, the mobility can increase even further and exceed 550 cm^2/Vs for sheet carrier densities of $1.5 \times 10^{11} cm^{-2}$. At lower densities, the background impurity scattering becomes prominent due to reduced shielding effect from the charged background impurity that lowers the 2DEG mobility.

Continuing the previous discussion, we propose mobility engineering in AlGaN channel devices by reducing the sheet charge density in the channel can significantly increase AlGaN channel mobility. However, simply reducing the channel sheet charge will severely limit the current density achievable in these devices. To increase the total sheet charge in the channel, it is therefore imperative to increase the number of parallel conducting channels in the devices which can be realized by using a multi-channel configuration. The combination of the low sheet charge per channel and a multi-channel configuration can simultaneously achieve high current density and high mobility. To retain good aspect ratio and thus good gate control in such tall devices, it is necessary to fabricate fin-shaped structures which can pinch off the channel charge by applying a lateral field by using a double-gate configuration, perpendicular to the c-plane as shown in Figure 3.5(b), instead of vertical field parallel to the c-plane, using a planar-gate configuration. The threshold voltage in such devices will be a strong function of the width of such fin structures, W. This strategy of gating by lateral depletion can also significantly reduce gate leakage current since dislocations, one of the key contributors to gate leakage current in GaN and AlGaN channel devices, are orientated in the c-plane and electrically insulated from the gate contact.

3.4. Summary

.   In conclusion, low-field transport properties in MOCVD-grown $Al_{0.70}Ga_{0.30}N/Al_{0.50}Ga_{0.50}N$ high electron mobility transistors have been conducted in this study. Low-resistance contacts were established on these devices using a reverse Al-composition graded contact scheme, whereby the Al-composition in the contact layer was progressively reduced from 70% to 30%, where it is significantly easier to establish metal-semiconductor ohmic contact and this resulted in a contact resistance of 2.1×10^{-5} $\Omega.cm^2$. A maximum room temperature mobility of 90 $cm^2/V \cdot s$ was obtained using effective mobility measurements. By considering each of the scattering components related to low-field transport in AlGaN films, the dominant scattering mechanism was found to be background impurity scattering for lower sheet charge density while at higher sheet charge densities, interface roughness limits the net mobility in the material. A careful analysis of the various scattering mechanisms also revealed potential pathway to significantly increase mobility in AlGaN channel transistors by utilizing multi-channel configuration with reduced charge per channel.

## CHAPTER 4

BREAKDOWN FIELD ENHANCEMENT BY EXTREME DIELECTRIC CONSTANT MATERIALS

4.1. BACKGROUND

As discussed earlier, various figures of merit suggest the excellent potential of AlGaN for future RF and power device technology. However, it remains an unsolved problem so far as to the approach of the actual critical electric field breakdown limit for lateral devices. It has only been achieved in vertical PN/PIN junctions previously[77, 78]. A non-uniform electric field distribution exists in the depletion region for a lateral device causing the electric field peaking. Tunneling breakdown at the Schottky gate or anode electrode also limits the breakdown further. Ultra-wide band gap semiconductors with average breakdown fields beyond an estimated value of 6 MV/cm are especially affected by such effects. AlGaN devices that employ metal-semiconductor based Schottky anodes or gates are limited by the breakdown strength of the Schottky diodes that typically have barrier heights of less than 2 eV. Critical breakdown electrical fields as high as 3.9 MV/cm have been demonstrated for $Al_{0.70}Ga_{0.30}N/Al_{0.50}Ga_{0.50}N$ high electron mobility transistors (HEMTs) while fields as high as 2.86 MV/cm have been demonstrated for Schottky gate electrodes for $Al_{0.70}Ga_{0.30}N$ metal semiconductor field effect transistors (MESFETs)[64, 79]. However, such numbers are still much lower than the estimated material breakdown field at these compositions (11 MV/cm for $Al_{0.70}Ga_{0.30}N$). Conventional low-k dielectrics such as Al_2O_3 and SiO_X that can be inserted between the gate or anode metal and semiconductor are not very good at offering significantly better performance since in these cases the device is limited by the breakdown of the metal-dielectric junction.

A potential solution to this issue could be to use high-k dielectric materials, such as Barium Titanate, $BaTiO_3$, which can be inserted between the semiconductor layers and the gate or anode metal and also inserted between the gate and drain region[80]. These materials could enable improved breakdown due to two reasons. First, a high permittivity dielectric constant layer significantly curtails the non-uniformity or peaking in lateral electric fields between drain to gate region for lateral transistors or cathode to anode for lateral diodes. Figure 4.1 shows the electric field and potential distribution profiles between the anode and cathode of a heterojunction diode in the semiconductor (channel) near the semiconductor-dielectric interface for two cases (inset of Figure 4.1). In one case a 25 nm low-k dielectric ($\varepsilon_r = 10$) is used while in another case a high k dielectric ($\varepsilon_r = 100$) is used for an applied anode to cathode voltage of 150 V and anode to cathode spacing of 200 nm. AlGaN is used as the channel material and the channel sheet charge density, n_s, is taken to be 1.5×10^{13} cm^{-2}. The use of high-k dielectric will result in a higher average breakdown field that will lead to a higher breakdown voltage due to a more uniformized E-field distribution. Secondly, the high k dielectric material below the anode or gate metal helps to significantly limit gate leakage related currents that ultimately reduces gate or anode leakage related breakdown (Figure 4.2).

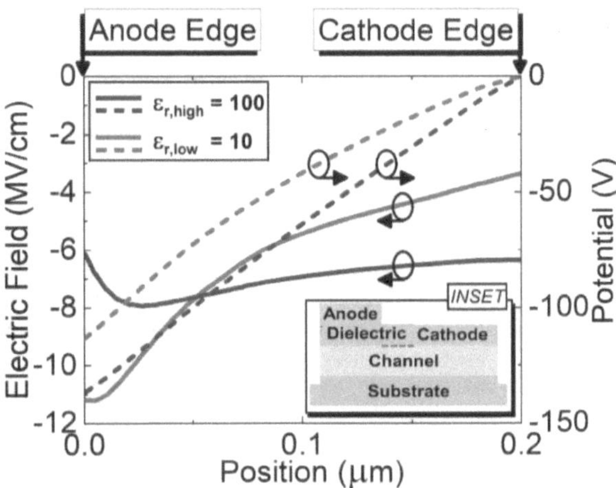

Figure 4.1 Total electric field and potential distribution comparison between cathode and anode at the dielectric-channel interface for a lateral heterojunction diode (inset) with 25 nm low permittivity dielectric materials (ε_r = 10) and high permittivity dielectric material (ε_r = 100) for an applied anode to cathode voltage, V_{AC} = -150 V. The dashed red line in the inset shows the cutline where the electric field and potential distribution have been simulated.

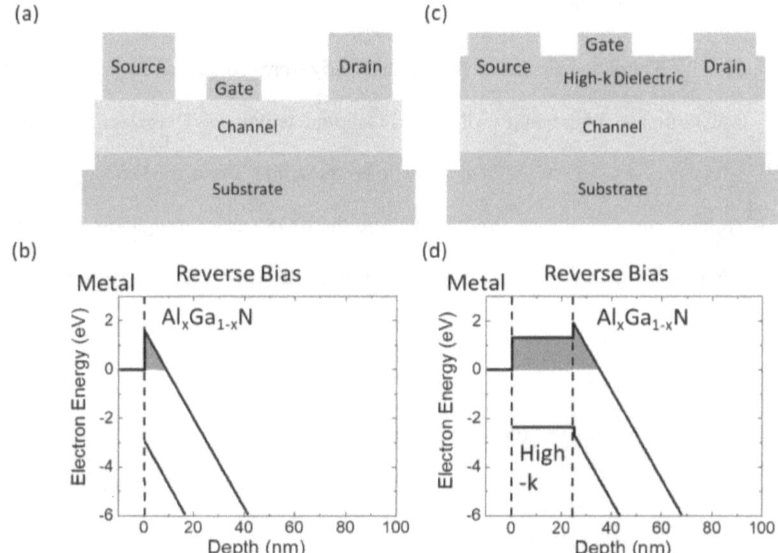

Figure 4.2 (a) and (c) Schematic illustrating a Schottky diode and a heterojunction diode with high permittivity dielectric material. (b) and (d) shows the corresponding simplified band diagram illustrating the electric fields in the devices under breakdown bias conditions.

4.2. HIGH BREAKDOWN BaTiO₃/AlGaN DIODES

4.2.1. EXPERIMENTAL DETAILS

The active channel layers for this study were grown via low-pressure (LP) MOCVD on 3 μm thick high quality AlN-(0001) sapphire templates. These template layers were also deposited by LPMOCVD at growth temperatures close to 1250 ℃. An atomic force microscope (AFM) was used to measure the RMS surface roughness of the active layers and the roughness was found to be 1.44 nm. The AFM scan of a portion of the sample is shown in Figure 4.3(a). The epilayer consisted of a 500 nm thick undoped i-$Al_{0.58}Ga_{0.42}N$ buffer layer grown pseudomorphically over a two-inch diameter AlN template at a growth temperature close to 1100 ℃. This was followed by the growth of a 60 nm thick [Si^+]-doped n-type $Al_{0.58}Ga_{0.42}N$ layer, with a doping concentration of 4×10^{18} cm^{-3} on top of the buffer layer as reported elsewhere[81]. High-resolution x-ray diffraction (XRD) spectra (BEDE D1 High Resolution Triple Axis X-ray Diffraction System) was used to determine the Al-composition of the grown film (Figure 4.3(b)).

Like the process of side contact formation in Section II, the ohmic contact was established to the MOCVD-grown n-doped AlGaN layer by selective area regrowth using molecular beam epitaxy (MBE). The SiO_2 hard mask was subsequently removed from the sample after the MBE contact regrowth process was completed. This was followed by a metal stack deposition of Ti/Al/Ni/Au [20/120/30/100 nm] on the regrown contact regions. The deposition was performed using an e-beam evaporation system to form ohmic contact metal stack. Device isolation was next performed by the use of inductively coupled plasma – reactive ion etching (ICP-RIE) plasma etching system with an RIE power of 30 W and a pressure of 5 mTorr. RF sputtering at 630 °C in an oxygen ambient was used to sputter $BaTiO_3$ on the sample. The dielectric constant of the $BaTiO_3$ films was estimated to be around 60. The dielectric constant of $BaTiO_3$ is a function of the grain size of the deposited film. For the deposition conditions used, the deposited film is expected to be either amorphous or nano-crystalline[82]. The observed dielectric constant of the deposited films are consistent with previous reports[83]. After the deposition of the $BaTiO_3$ was completed, the deposited $BaTiO_3$ was removed completely using an SF_6 plasma ICP-RIE etching technique for the control Schottky barrier diodes. However, for the heterojunction diode $BaTiO_3$ was selectively etched away only from the contact regions. Anodes were finally deposited by e-beam evaporation of Pt/Au [60/100 nm] metal stack via e-beam evaporation for both the Schottky barrier diode (control sample) and the BTO/AlGaN heterojunction diodes. Figure 4.4(a) show a schematic representation of the fabricated heterojunction diode while the equilibrium energy-band diagrams calculated using self-consistent one-dimensional Poisson-Schrodinger solver, under the anode, for the $BaTiO_3$ sample is shown in Figure 4.4(b). The conduction band offset and Schottky barrier height was estimated from known electron affinity and work function values for the materials[84, 85]. The schematic and energy band-diagram of the control (Schottky) device, under the anode, are shown in Figures 4.4(c) and (d), respectively.

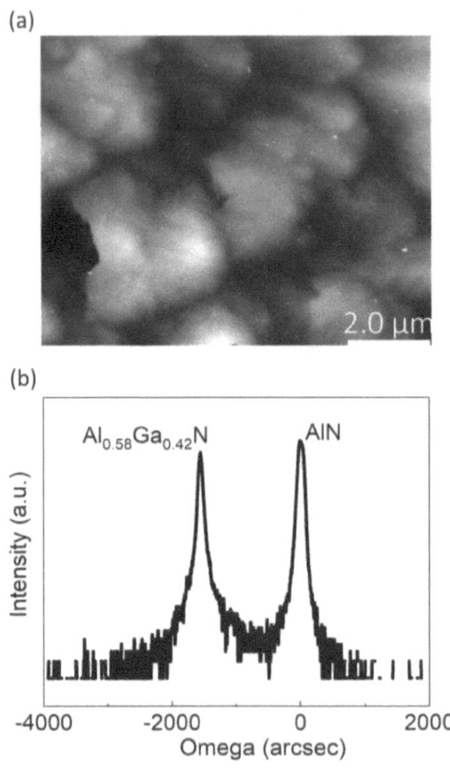

Figure 4.3 (a) Surface AFM scan of the sample showed an RMS surface roughness of 1.44 nm and (b) XRD used to confirm the composition of the MOCVD grown film

Four-terminal ungated van der Pauw (VDP) structures without $BaTiO_3$ were used for Hall measurements. Based on the Hall measurements, the sheet resistance was found to be 6.2 kΩ/$\square$. The Hall mobility and sheet charge density were found to be 65 cm^2/V·s and 1.55×10^{13} cm^{-2}, respectively. An Agilent B1500 parameter analyzer was used to perform electrical characterization of the devices. Capacitance voltage measurements performed on a 16 μm×100 μm BTO/AlGaN diode found the undepleted sheet charge density to be 1.06×10^{13} cm^{-2}. The measured and simulated charge density are shown in Figure 4.5. The extracted doping density from the capacitance voltage measurement was 4×10^{18} cm^{-3}, as expected. Figure 4.6(a) shows the forward current voltage behavior of a control Schottky diode and a BTO/AlGaN heterojunction diode with an anode to cathode spacing of 700 nm. The turn-on voltage of the Schottky diode and the heterojunction diode are 0.9 V and 1.5 V, respectively. Both these devices display significantly lower turn-on voltage compared to what is typically observed for AlGaN PN/PIN diodes. A differential on-resistance (R_{ON}), of 31 mΩ.cm^2 was measured for the Schottky diodes, while an R_{ON} of 56 mΩ.cm^2 was measured for the heterojunction diodes. The increase of R_{ON} in the heterojunction device is likely due to the presence of $BaTiO_3$ layer which is nominally undoped.

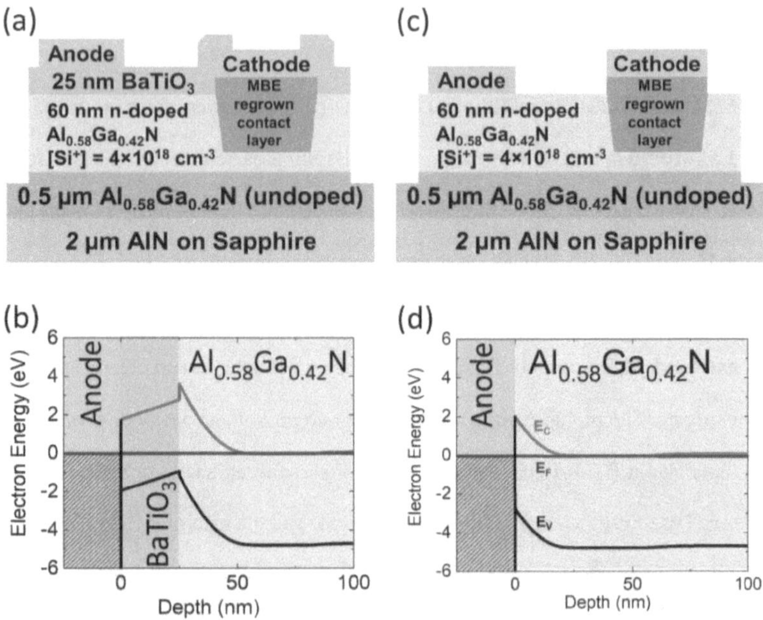

Figure 4.4 (a). Schematic of the lateral Pt/BaTiO₃/Al₀.₅₈Ga₀.₄₂N heterojunction diode ($\varepsilon_{r,BaTiO3}$ = 100), (b) band-diagram under the anode for the heterojunction diode (c) schematic of the fabricated lateral Pt/Al₀.₅₈Ga₀.₄₂N Schottky diode, and (d) band-diagram under the anode for the Al₀.₅₈Ga₀.₄₂N Schottky diode

Several randomly selected Schottky and BTO/AlGaN diodes were used for electrical breakdown performance evaluation and comparison. Figure 4.6(b) shows the breakdown characteristics for a Schottky diode and a heterojunction diode, with anode to cathode spacing of 0.19 μm and 0.18 μm respectively. While the Schottky diode breaks down at an anode-to-cathode voltage of 42 V, the heterojunction diode breaks down at a much higher 155 V, despite both devices having similar anode to cathode spacing. The improvement in the average breakdown field is attributed to a reduced peak field and the suppression of anode leakage current due to the $BaTiO_3$ layer. Assuming total depleted sheet charge density of 1.55×10^{13} cm^{-2} the peak vertical field was estimated to be 3.15 MV/cm. The average lateral electric field of the Schottky barrier diode was estimated to be 1.58 MV/cm (assuming a pinch-off voltage of -12 V), which gives a total average electric field of 3.5 MV/cm. The average lateral field and the peak vertical field of the BTO/AlGaN diode were estimated to be 7.94 MV/cm and 3.15 MV/cm respectively. This gave an average total breakdown electric field of 8.5 MV/cm. Multiple randomly chosen devices were measured for both the Schottky diodes and the heterojunction diodes and the results are shown in Figure 4.7(a). In general, for the heterojunction diodes, for an anode to cathode spacing in the range of 0.2 – 0.3 μm, the average breakdown fields observed were in the range of 6-8 MV/cm. The highest reported average breakdown fields for various semiconductor materials as a function of the bandgaps are shown in Figure 4.7(b). The value reported in this work, 8.5 MV/cm, is the highest reported experimental average breakdown field for any semiconductor material to date.

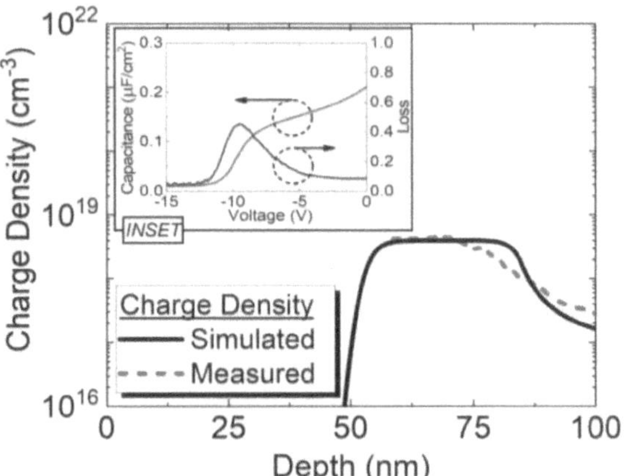

Figure 4.5 Comparison of the measured charge profile to the simulated charge profile of the BaTiO$_3$/Al$_{0.58}$Ga$_{0.42}$N diode; (INSET) capacitance-voltage measurement results from BaTiO$_3$/Al$_{0.58}$Ga$_{0.42}$N diodes

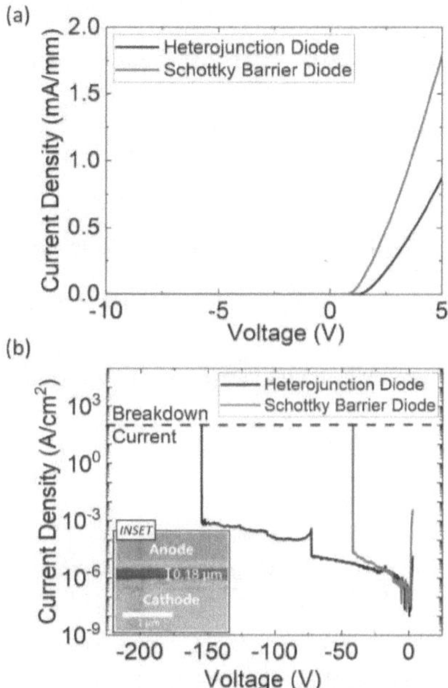

Figure 4.6 Comparison between Schottky barrier diode and BaTiO$_3$/Al$_{0.58}$Ga$_{0.42}$N heterojunction diode (a) forward IV characteristics and (b) breakdown performance; (INSET) anode to cathode spacing for the heterojunction diode with V_{BR} = -155 V, as measured using a scanning electron microscope (SEM)

For the estimation of the vertical electric fields, we used Gauss's Law. The peak vertical field arises due to the ionized donors in the channel given by

$$E_{vertical} = \frac{eN_D x_D}{\epsilon_s} = \frac{e \cdot n_s}{\epsilon_s}$$

where e is the elementary electric charge, n_s is the sheet charge density in an ungated, undepleted structure without $BaTiO_3$ and ε_s is the dielectric constant of the semiconductor.

The average horizontal electric field arises due to the voltage applied between the two electrodes and should be horizontal in direction. The peak lateral field at the gate edge is higher than the average lateral field

$$E_{lateral} = \frac{V_{app} - V_{depletion}}{L_{AC}}$$

Therefore, we used the average field as a lower bound to estimate the maximum field in the structure as

$$E_{total} = \sqrt{(E_{vertical})^2 + (E_{lateral})^2}$$

where V_{app} is the voltage applied between anode and cathode, $V_{depletion}$ is the lateral depletion voltage and L_{AC} is the spacing between the anode and cathode as measured using a scanning electron microscope (SEM). The results of the calculated electric fields for the $BaTiO_3$/AlGaN diodes and Schottky diodes are shown in Table 4.1 and 4.2, respectively.

Calculations of electric fields for the heterojunction diode are shown at the table below:

Table 4.1 Calculation of electric fields for the BTO/AlGaN heterojunction diodes

V_{BR} (V)	L_{AC} (cm)	$E_{vertical}$ (MV/cm)	$E_{lateral}$ (MV/cm)	$E_{average}$ (MV/cm)
184	2.98×10^{-05}	3.15	5.78	6.58
138	2.25×10^{-05}	3.15	5.60	6.42
155	1.80×10^{-05}	3.15	7.94	8.55
144	1.79×10^{-05}	3.15	7.37	8.02

Calculations of electric fields for the Schottky diodes are shown below:

Table 4.2 Calculation of electric fields for the Schottky diodes

V_{BR} (V)	L_{AC} (cm)	$E_{vertical}$ (MV/cm)	$E_{lateral}$ (MV/cm)	$E_{average}$ (MV/cm)
90	2.95×10^{-05}	3.15	2.64	4.11
43	3.00×10^{-05}	3.15	1.03	3.31
42	1.90×10^{-05}	3.15	1.58	3.52

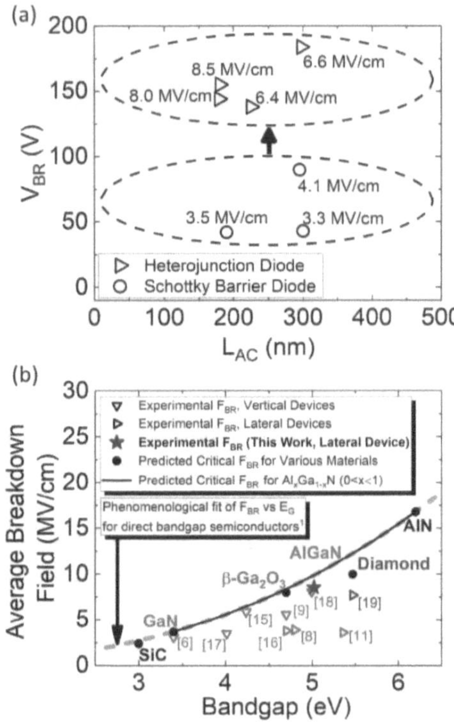

Figure 4.7 (a) Breakdown performance for lateral $Pt/BaTiO_3/Al_{0.58}Ga_{0.42}N$ heterojunction diodes and AlGaN Schottky barrier devices for L_{AC} < 500 nm and (b) comparison of experimental average breakdown fields in various semiconductor devices achieved to date (**pink**: GaN reports, **green**: Ga_2O_3 reports, **red**: AlGaN reports, **brown**: diamond reports)[63, 77-79, 86-89]. The dashed orange line shows the phenomenological fit of critical breakdown field (F_{BR}) as a function of bandgap (E_G) for direct bandgap semiconductors. The solid black circles represent the predicted critical breakdown fields for various materials[47, 78, 87, 90, 91].

We have performed TCAD simulation for both AlGaN Schottky device and the BTO/AlGaN heterojunction device at the breakdown bias conditions. Both devices were assumed to have an anode to cathode spacing of 200 nm. It should be noted that simulations can overestimate the peak fields since they assume angular geometries. For instance, rectangular structures with 90° corners were used as anode and cathode contacts. However, in real devices the metal corners will be more sloped and rounded and hence the simulated electric fields near the metal corners will be overestimated. Moreover, simulations do not include the effects of surface and interface charges. Hence, although simulations may provide a good comparison between the BTO/AlGaN device and the Schottky AlGaN device and can be used to isolate the likely breakdown locations, they should not be used to extract exact values of the peak electric fields at breakdown in this case.

The results of the simulations are shown below in Figure 4.8 and 4.9. The average breakdown electric fields in the case of the Schottky diodes, estimated both from the experimental measurement and from the TCAD simulation, are > 4 MV/cm, which is typically considered as the electric field required for the onset of Fowler-Nordheim tunneling[92, 93]. The breakdown of the Schottky diodes is most likely being limited by the tunneling related breakdown as previously discussed. A cutline represented by a red dashed line between the anode to cathode in Figure 4.8 shows that there is a sharp electric field peak exceeding 4 MV/cm at the interface of the metal semiconductor junction. This is where the tunneling breakdown most likely occurs.

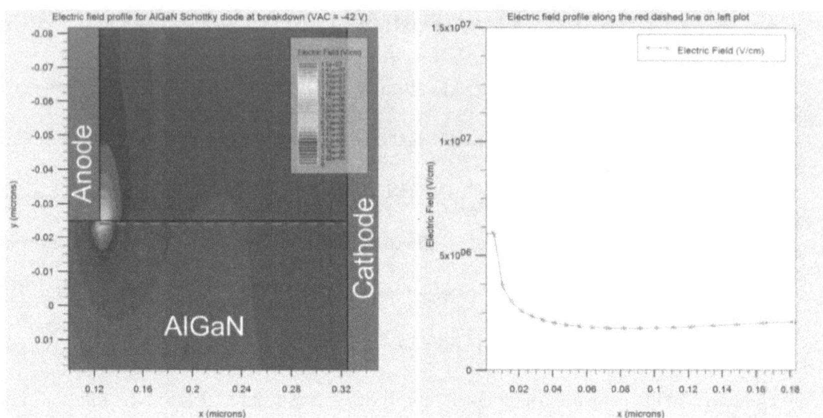

Figure 4.8 TCAD electric field profile for AlGaN Schottky diode (left) between anode and cathode at breakdown condition (V_{AC} = -42 V) and (right) along the red dashed line

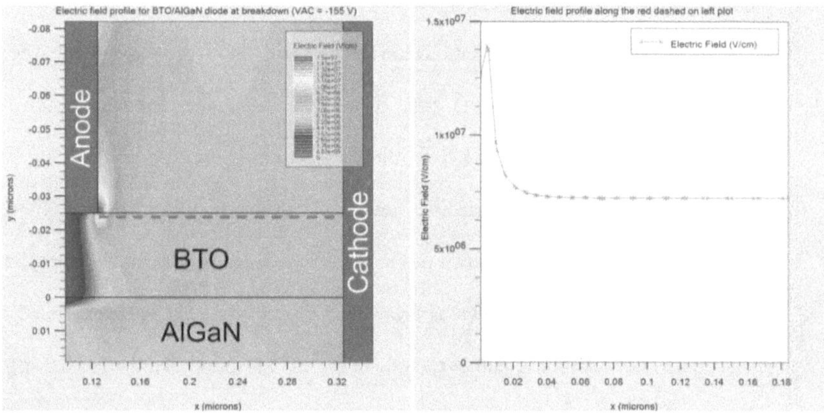

Figure 4.9 TCAD electric field profile for BTO/AlGaN diode (left) between anode and cathode at breakdown condition (V_{AC} = -155 V) and (right) along the red dashed line

For the case of the BTO/AlGaN diode, however, the breakdown is most likely not due to Fowler-Nordheim tunneling since the breakdown happens at a significantly higher field than 4 MV/cm, as shown by the cutline represented by red dashed line in Figure 4.9. For the BTO/AlGaN structure discussed in this chapter, the high permittivity dielectric between the anode and semiconductor effectively reduces the vertical field by corresponding charge screening. However, the negative charge present at the gate edge towards drain creates a large lateral electric field which is not screened by the high permittivity dielectric. Therefore, the metal-$BaTiO_3$ interface is most likely the location of the breakdown of the breakdown in the heterojunction diode. Previous work on vertical $BaTiO_3/\beta\text{-}Ga_2O_3$ diodes [94], have provided some evidence of a trap assisted tunneling (Poole-Frenkel emission) related breakdown which might also be the breakdown mechanism here. However, further studies are needed to fully understand the breakdown mechanisms in these diodes.

4.3. High Breakdown BaTiO$_3$/AlN/GaN Transistors

As discussed previously, a high-k dielectric such as BaTiO$_3$ can be integrated into lateral devices to significantly increase the breakdown performance. This is achieved by reducing the electric field peaking and the reducing the gate leakage current. The breakdown performance improvement is expected to be more significant in high charge devices [80]. AlN/GaN HEMTs possess extremely high sheet charge density exceeding 2.5×10^{13} cm^{-2} due to the high positive polarization charge present at the interface of AlN and GaN and are ideal candidates for the demonstration of the breakdown performance improvement due to the integration of BaTiO$_3$ dielectric layer. It is known from previous literature these devices show poor breakdown performance (V$_{Breakdown}$ ~ 20 V), even with the integration of low-k dielectrics such as Al$_2$O$_3$.

The schematic of the AlN/GaN HEMT structure and the corresponding band diagram is shown in Figure 4.10 (a) and (b) respectively. The structures were grown via MBE. Selective area contact regrowth of n$^+$ GaN was performed to obtain low-resistance contact. 25 nm BaTiO$_3$ was then deposited using RF sputtering using similar conditions as before. This was wet-etching of the BaTiO$_3$ followed by e-beam evaporated non-alloyed Ti/Al/Ni/Au metal stack for source/drain contact. This was followed by the deposition of Ni/Au to act as gate contact. Transfer length method showed a contact resistance of 0.26 Ohm.mm. Hall measurement showed a sheet charge density around 2.6×10^{13} cm^{-2} and a hall mobility of around 776 cm^2/V.s. Three-terminal output and transfer characteristics were performed on transistors with L$_G$ = 0.6 μm, L$_{GS}$ = 0.2 μm and L$_{GD}$ = 1.4 μm which showed a maximum current density of 1.83 A/mm (Figure 4.10 (c)) and a maximum transconductance of 0.254 S/mm (Figure 4.10 (d)). Breakdown measurement was performed and a breakdown voltage, V_{GD} of 58 V was observed (Figure 4.10 (e)). The This is significantly higher than what has been previously reported in the literature. Although this preliminary work is very encouraging, the gate leakage current was found to high, which likely limited the breakdown characteristics. Work is currently in progress to lower the gate leakage current and push the breakdown voltage beyond 100 V.

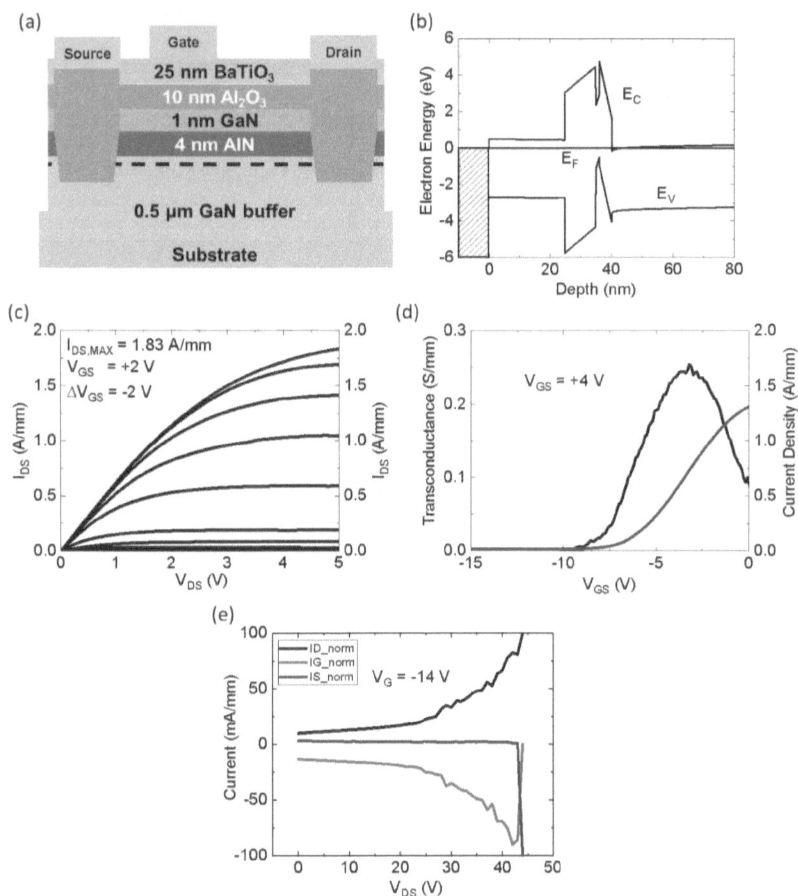

Figure 4.10 (a) Schematic of the AlN/GaN HEMT with BaTiO₃ dielectric layer, (b) under the gate band-diagram, (c) output characteristics, (d) transfer characteristics and (e) breakdown characteristics of the BaTiO₃/AlN/GaN HEMT

A key benefit of using $BaTiO_3$ as a gate or anode dielectric is that it enables the fabrication of ultra-scaled devices while meeting breakdown voltage requirements. This is particularly beneficial for low mobility materials like AlGaN and Ga_2O_3. These materials display higher sheet resistance due to their lower mobility. To enable high frequency operation, it is necessary to scale these devices fabricated using these materials. The presence of the high-k dielectric also leads to scaling of the gate to channel distance which can result in lower output conductance and higher transconductance[80]. However, it should be noted that in such a heterojunction with a high-k dielectric, the drain depletion region is longer than the case with a low-k dielectric and correspondingly higher gate to channel capacitance due to the high-permittivity region between the gate and drain[80]. As a result, the peak cutoff frequency is expected to be lower than a device with low-k barrier. It is therefore important to properly optimize the thickness of the high-permittivity region to manage the tradeoff between field management and gate–drain capacitance dielectric.

4.4. SUMMARY

In summary. this chapter discussed one of the most important problems in ultra-wide bandgap semiconductors whereby approaching the actual critical electric field breakdown limit in such materials, with breakdown electric field > 6 MV/cm, cannot be achieved in lateral devices. This is due to a) the non-uniform electric field distribution in the depletion region for a lateral device (which causes electric field peaking) and b) the bottleneck of tunneling related breakdowns at the Schottky gate or anode electrode. A novel solution was devised and demonstrated for a heterojunction diode where an extreme permittivity material, $BaTiO_3$, was inserted between the metal and semiconductor layers and between anode and cathode region. The peak field at the anode edge was significantly reduced by employing this strategy and an average breakdown field exceeding 8 MV/cm was achieved for devices with anode to cathode spacing of <0.2 μm. The use of a high-k dielectric can thus more effectively utilize the high breakdown fields in ultra-wide bandgap materials by proper electric field management.

CHAPTER 5:

3D FINFET HIGH AL-COMPOSITION ALGAN CHANNEL

TRANSISTORS

5.1. BACKGROUND

While both MOCVD grown contact layer approach and MBE regrown contact layers on MOCVD grown channel approach device growth have been unable to produce sub-1 Ω.mm contact resistance required for high performance transistors, it is possible to improve the device on-resistance by manipulating device geometry. The modified device geometry is shown in Figure 5.1. The channel is divided into multiple conducting and non-conducting regions. For such a structure with two-terminals, the total device resistance is given by:

$$R_T = \frac{2 \cdot R_{contact}}{W_{contact}} + \frac{R_{sheet} \cdot L}{n \cdot W_s'}$$

where $R_{contact}$, $W_{contact}$, R_{sheet}, n, $W_{S'}$, L and R_T are the contact resistance, contact width, sheet resistance, number of active stripes, width of one active stripe and the channel length respectively. Higher contact resistance is undesirable in a device since this will result in a non-negligible voltage drop across the contacts during normal device operation. As a result, the effective voltage drop across the device channel will be lowered. To improve the normalized on-resistance, the effective voltage drop across the channel needs to be increased, i.e. the fractional voltage drop across the channel needs to be increased.

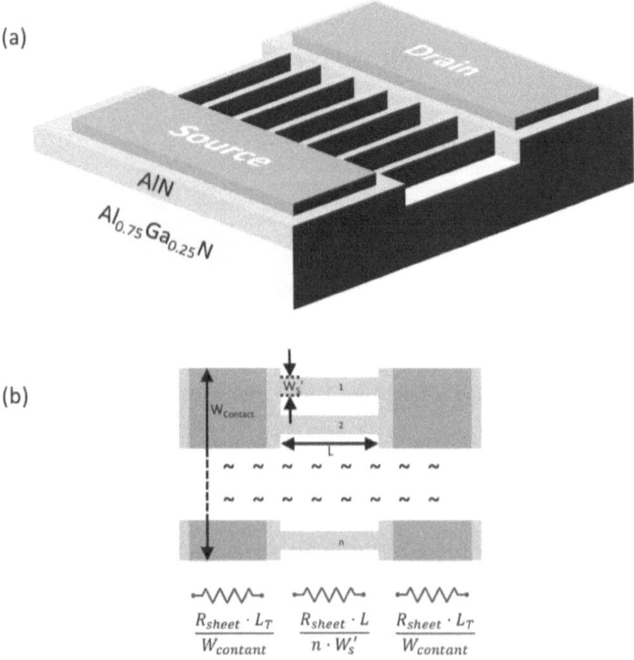

Figure 5.1 (a) 3D and (b) plan-view schematic of a typical FinFET showing the effect on effective resistance using a larger periphery with fin-shaped channels.

The fractional voltage drop across the channel is given by series voltage divider rule:

$$\Rightarrow V_{channel,fractional} = \frac{\frac{R_{sheet} \cdot L}{n \cdot W_s'}}{\left(\frac{2 \cdot R_{contact}}{W_{contact}} + \frac{R_{sheet} \cdot L}{n \cdot W_s'}\right)}$$

$$\Rightarrow V_{channel,fractional} = \frac{1}{\left(\frac{2 \cdot R_{contact}}{W_{contact}} \cdot \frac{n \cdot W_s'}{R_{sheet} \cdot L} + 1\right)}$$

We can denote the effective width of the channel, $W_{channel}$ as the product of the total number of active stripes and the width of one active stripe.

$$\Rightarrow V_{channel,fractional} = \frac{1}{\left(\frac{2 \cdot R_{contact}}{R_{sheet} \cdot L} \cdot \frac{W_{channel}}{W_{contact}} + 1\right)}$$

$$\Rightarrow V_{channel,fractional} = \frac{1}{\left(1 + \frac{2 \cdot R_{contact}}{R_{sheet} \cdot L} \cdot x\right)} ; \ 0 \leq x \leq 1$$

where x denotes the ratio between the width of the active region of the channel to the width of the contact region. As the channel width is reduced x will progressively become smaller than 1 and the fractional voltage drop in the channel will increase. As a result, the current for the same applied voltage will be higher as the average electric field, $\left(\xi_{average} = \frac{V_{channel}}{L}\right)$, in the channel will be higher. This should result in higher current density as the current density in the channel, $J_{channel}$, is given by

$$J_{channel} = e \cdot n_s \cdot \mu \cdot \xi_{average}$$

The on-resistance normalized to the effective channel width will be correspondingly lower as can be seen by the equation below:

$$\Rightarrow R_T \cdot n \cdot W_S' = 2 \cdot R_{contact} \cdot \left(\frac{n \cdot W_S'}{W_{contact}}\right) + R_{sheet} \cdot L$$

$$\Rightarrow R_T \cdot W_{channel} = 2 \cdot R_{contact} \cdot x + R_{sheet} \cdot L; \ 0 \le x \le 1$$

Using the above equations, we can plot the fractional voltage drop in the channel (Figure 5.2(a) and (b)) and the normalized on-resistance (Figure 5.2(c) and (d)) as a function of the ratio of total channel width to contact width.

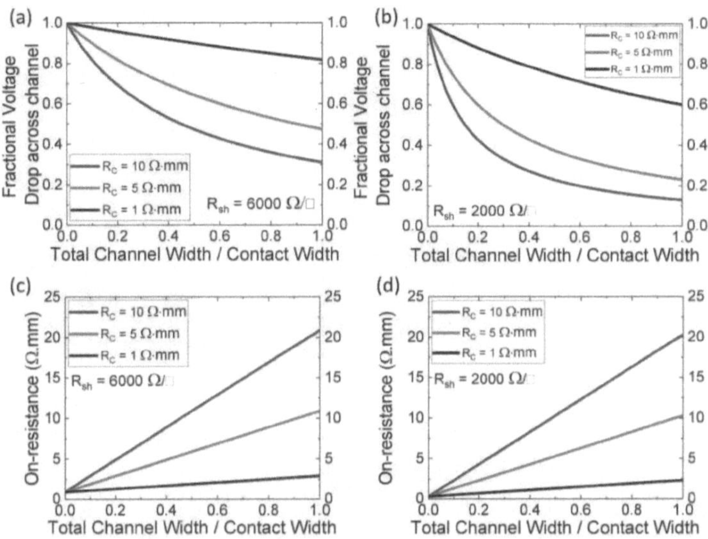

Figure 5.2 Fractional voltage drop across the channel for various contact resistances for sheet resistance of (a) 6000 $\Omega/\square$ and (b) 2000 $\Omega/\square$ as a function of the ratio between total channel and the contact width. On-resistance for various contact resistances for sheet resistance of (a) 6000 $\Omega/\square$ and (b) 2000 $\Omega/\square$ as a function of the ratio between total channel and the contact width

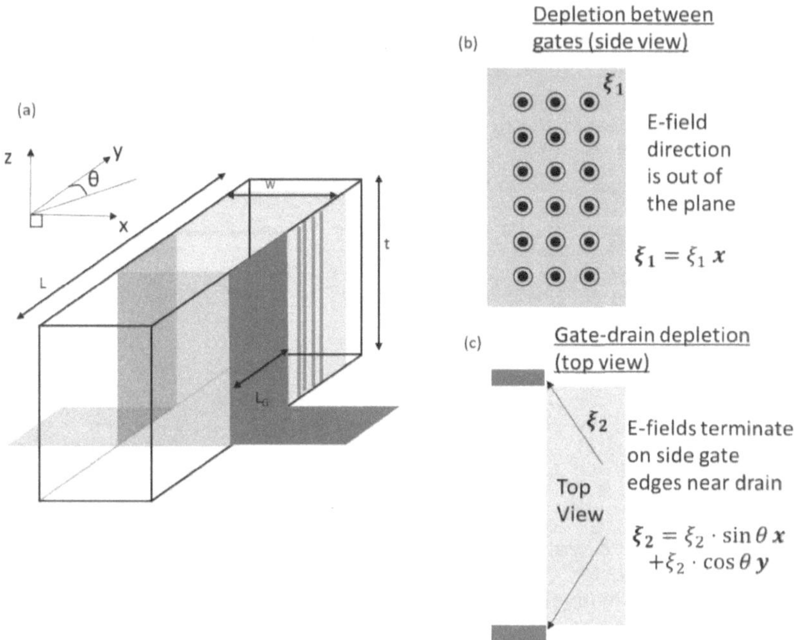

Figure 5.3 (a) FinFET model showing 3D schematic of a typical fin with the depletion regions between the gate fingers (green shaded) and between gate and drain (orange shaded). (b) Electric field due to the depletion region between the gate fingers and (c) due to the depletion region between gate and drain.

An analytical model for the electric-field distribution has been developed for a FinFET. A 3D schematic of such a fin is shown in Figure 5.3(a). The depletion regions in the fin can be assumed to be separated into two distinct regions. Region I, shaded in green, is the depletion region between the gate metals on each side of the fins while region II,

105

shaded in orange, is the depletion region between the gate and the drain region. Figure 5.3(b) shows the electric field directions for the depletion region between the gates, which shows that the fields will be parallel to the x direction only, as the fields are terminated in the gate metals. In contrast, as Figure 5.3(c) shows, the depletion charges between the gate and the drain region will have field termination primarily happening the gate edge towards drain and will have both an x- and y-component.

To estimate the electric fields due to depletion region between the gates, we can use Gauss's law as shown below:

$$\oint \xi \cdot dS = \frac{e \cdot n_{s,side} \cdot A}{\epsilon_s}$$

$$\Rightarrow \xi \cdot 2A = \frac{e \cdot n_{s,side} \cdot A}{\epsilon_s}$$

$$\Rightarrow \xi = \frac{e \cdot n_{s,side}}{2\epsilon_s} = \frac{e \cdot n_{3d}}{\epsilon_s} \cdot \frac{W}{2}$$

$$\therefore \xi(x) = \left(\frac{e \cdot n_{3d}}{\epsilon_s} \cdot \frac{W}{2} \right) \cdot \left(\frac{x}{W/2} \right), assuming\ x = 0\ at\ fin\ center$$

where ξ is the electric field due to the region I depletion charge, $n_{s,side}$ is the sheet charge looking from the side of the fins, ε_s is the dielectric constant of fin material, W is the width of the fins and n_{3D} is the volumetric charge in the fins.

Next, to estimate the electric fields due to depletion region II, the depletion region can be visualized as series of sheets of charges in the y-z plane which in turn are formed by lines of charges along the z-direction. The electric fields can thus be found by summing up the E-field due to lines of charge along the y-direction followed by summing sheets of charge along the x-direction. Using the equations below, expressions for the total E-field can be found. First the electric-field along the y-direction is calculated:

$$d\xi_y = \frac{q n_{3d}\, dx'\, dy' \cdot \cos(\theta)}{2\pi\epsilon_s \sqrt{(x-x')^2 + (y-y')^2}}$$

$$\xi_y = \int_{y=0}^{L_{Dep}} \int_{x=-W/2}^{W/2} \frac{q n_{3d}(y-y')\, dx'\, dy'}{2\pi\epsilon_s [(x-x')^2 + (y-y')^2]}$$

$$
\begin{aligned}
\xi_y ={}& \frac{q n_{3d}}{4\pi\epsilon_s}\Bigg[\left\{ \left(W/2 - x\right)\cdot\ln\frac{\left(W/2 - x\right)^2 + y^2}{\left(W/2 - x\right)^2 + (y - L_{Dep})^2} - \left(-W/2 - x\right)\ln\frac{\left(-W/2 - x\right)^2 + y^2}{\left(-W/2 - x\right)^2 + (y - L_{Dep})^2} \right\} \\
&- \left\{ 2(y - L_{Dep})\arctan\left(\frac{W/2 - x}{y - L_{Dep}}\right) - 2(y - L_{dep})\arctan\left(\frac{-W/2 - x}{y - L_{Dep}}\right) \right\} \\
&+ \left\{ 2y\cdot\arctan\left(\frac{W/2 - x}{y}\right) - 2y\cdot\arctan\left(\frac{-W/2 - x}{y}\right) \right\} \Bigg]
\end{aligned}
$$

Next, electric fields in the x-direction is also calculated.

$$d\xi_x = \frac{qn_{3d}\,dx'dy' \cdot \sin(\theta)}{2\pi\epsilon_s\sqrt{(x-x')^2+(y-y')^2}}$$

$$\xi_x = \int_{y=0}^{L_{Dep}} \int_{x=-W/2}^{W/2} \frac{qn_{3d}(x-x')dx'dy'}{2\pi\epsilon_s[(x-x')^2+(y-y')^2]}$$

$$\begin{aligned}
\xi_x &= \frac{qn_{3d}}{4\pi\epsilon_s}\left[\left\{(L_{Dep}-y)\cdot\ln\frac{(L_{Dep}-y)^2+\left(x+W/2\right)^2}{(L_{Dep}-y)^2+\left(x-W/2\right)^2} - (-y)\ln\frac{(-y)^2+\left(x+W/2\right)^2}{(-y)^2+\left(x-W/2\right)^2}\right\}\right. \\
&\quad -\left\{2\left(x-W/2\right)\arctan\left(\frac{L_{Dep}-y}{x-W/2}\right) - 2\left(x-W/2\right)\arctan\left(\frac{-y}{x-W/2}\right)\right\} \\
&\quad \left. +\left\{2\left(x+W/2\right)\cdot\arctan\left(\frac{L_{Dep}-y}{x+W/2}\right) - 2\left(x+W/2\right)\cdot\arctan\left(\frac{-y}{x+W/2}\right)\right\}\right]
\end{aligned}$$

The above equations can be used to calculate the electric field distributions along the fins between the gate to drain regions. An example 3-D distribution of electric field for a FinFET with a fin width and height of 300 nm with a top sheet charge of 3×10^{13} cm^{-2} is shown in Figure 5.4. As can be seen, the peak electric field occurs at the two corners of the gate edge near drain. For this scenario, the peak field is 6.75 MV/cm. The peak electric field in the device for a specific bias condition will determine the breakdown condition for the device.

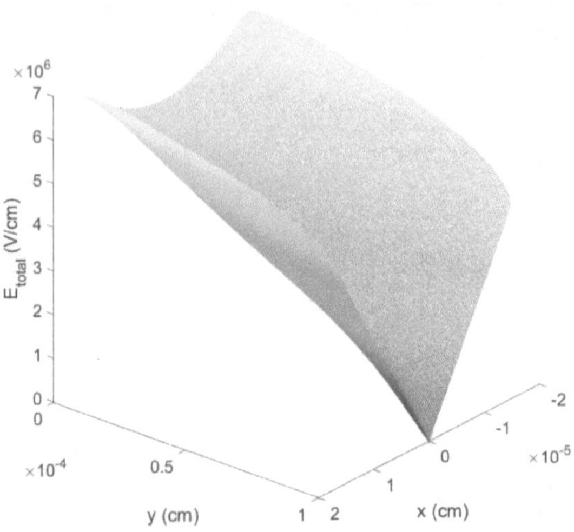

Figure 5.4 Total electric field distribution for a FinFET with fin width of 300 nm and fin height of 300 nm with a top sheet charge of 3×10^{13} cm^{-2}

Based on these equations, it is possible to predict various performance parameters for FinFETs and correspondingly compare them with similar performance parameters expected from a planar device. The first step is to determine the bias conditions that will produce a total electric field equal to the critical breakdown electric field in the material. Using the same bias conditions, the electric field in the y-direction can then be determined which when integrated along the y-direction will produce the maximum gate to drain voltage that can be sustained in the device prior to breakdown. The maximum output power from the FinFET can be estimated from the class-A amplifier equation given by

$$P_{out} = \frac{I_{MAX} \times V_{BR}}{8}$$

where I_{max} is the maximum current density and can be estimated by total sheet charge times the saturated electron velocity and the V_{BR} is the breakdown gate to drain bias. We can then determine the RF performance of these FinFETs. To illustrate this with an example, we have chosen the parameters shown in Table 5.1.

Table 5.1 Parameters used to estimate the output power density and the RF performance of FinFETs

Params	Value
$\xi_{critical}$	6 MV/cm
v_{sat}	1.5×10^7 cm/s
R_0	10 Ohm.mm
R_{in}	200 Ohm/mm
W_{FIN}	100 nm
t_{FIN}	300 nm

To evaluate the RF performance, unity current gain cutoff frequency, f_T, unilateral power gain cutoff frequency, f_{MAX}, power gain, G_P, and the operating frequency, f_{OP}, were then calculated as follows:

$$f_T = \frac{v_{sat}}{2\pi(L_G + L_{GD})}$$

$$f_{MAX} = \frac{f_T}{2}\sqrt{\frac{R_0}{R_{in}}}\frac{1}{W}$$

$$G_P = \left(\frac{f_{MAX}}{f}\right)^2$$

where v_{sat} is the saturated electron velocity, L_G is the gate length, L_{GD} is the gate to drain spacing, R_0 is the output resistance, R_{in} is the input resistance, W is the device width and P_{OUT} is the output power. Tables 5.2 and 5.3 shows the RF performance for FinFET devices for gate lengths of 100 and 250 nm, respectively. Similarly, Tables 5.4 and 5.5 shows the expected performance for planar devices with similar material and electrical properties. It should be noted that the current density is assumed to be the product of sheet charge and saturated electron velocity. Although, the equations estimate the performance to a first order and ignore many second order effects, the estimated numbers do provide a reasonable comparison between the two types of devices. Based on the tables it is apparent that significant RF performance improvement can be expected by moving from a planar device to a FinFET device configuration.

Table 5.2 RF performance estimation of FinFETs with parameters provided in Table 5.1

with a gate length, L_G, of 100 nm

L_G = 100 nm					
L_{GD} (nm)	f_T (GHz)	f_{MAX} (GHz)	f_{OP} (GHz) (G_P = 10 dB)	f_{OP} (GHz) (G_P = 15 dB)	P_{OUT} (W/mm)
100	119.4	298.4	94.4	53.1	74.2
250	68.2	170.5	53.9	30.3	164.5
500	39.7	99.5	31.5	17.7	294

Table 5.3 RF performance estimation of a FinFET with parameters provided in Table 5.1

with a gate length, L_G, of 250 nm

L_G = 250 nm					
L_{GD} (nm)	f_T (GHz)	f_{MAX} (GHz)	f_{OP} (GHz) (G_P = 10 dB)	f_{OP} (GHz) (G_P = 15 dB)	P_{OUT} (W/mm)
100	68.2	170.5	53.9	30.3	74.2
250	47.7	119.4	37.7	21.2	164.5
500	31.8	79.6	25.2	14.2	294

Table 5.4 RF performance estimation of planar devices with parameters provided in Table 5.1 with a gate length, L_G, of 100 nm

$L_G = 100$ nm					
L_{GD} (nm)	f_T (GHz)	f_{MAX} (GHz)	f_{OP} (GHz) ($G_P = 10$ dB)	f_{OP} (GHz) ($G_P = 15$ dB)	P_{OUT} (W/mm)
100	119.4	298.4	94.4	53.1	14.9
250	68.2	170.5	53.9	30.3	37.7
500	39.8	99.5	31.5	17.7	71.6

Table 5.5 RF performance estimation of a planar device with parameters provided in Table 5.1 with a gate length, L_G, of 250 nm

$L_G = 250$ nm					
L_{GD} (nm)	f_T (GHz)	f_{MAX} (GHz)	f_{OP} (GHz) ($G_P = 10$ dB)	f_{OP} (GHz) ($G_P = 15$ dB)	P_{OUT} (W/mm)
100	68.2	170.5	53.9	30.3	14.9
250	47.7	119.4	37.7	21.2	37.7
500	31.8	79.6	25.2	14.2	71.6

5.3. FinFET Process Development

A FinFET is a multi-gate device where the channel charge is located in distinct "Fin"-shaped islands and surrounded by the gate electrode at least from two directions (double gate FET) and in some cases from three directions (top and sides; also known as trigate FET). Due to such a geometry where the channel E-field termination can happen over a larger region, these devices typically show better pinch-off behavior compared to planar FETs which can only have field terminations at the top gate electrode. Besides this, as already discussed in the introductory section, these structures have the added advantage of significantly reducing the detrimental effects of high contact resistance. All the FinFET structures described in this book has been fabricated using optical lithography. Using a controlled over-exposure of the patterns, sub-wavelength features have been achieved, which negated the necessity of using complicated e-beam lithography-based approaches. This not only reduced the process improvement turnaround time but also increased the process yield. In the next section, the full process development and optimization is described.

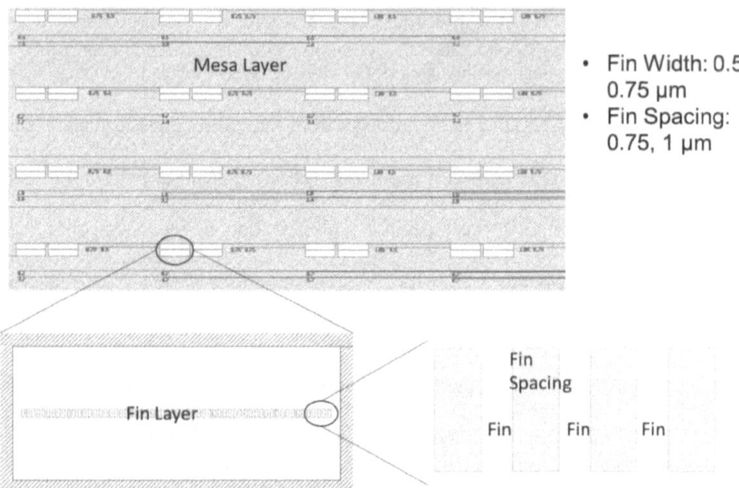

Figure 5.5 Layout of the mask used for fabricating the fins. The fins lie between the source and drain regions and various spacings, ranging from 0.75 to 1 μm, have been used to obtain the best possible fin shapes.

The mask layout used for these studies are shown in Figure 5.5. Photolithography for the reproduction of the fin structures on the photoresist can be performed together with the mesa photolithography process since the etch depth and chemistry will be the same for the structures. It is also possible to expose the fin structures separately.

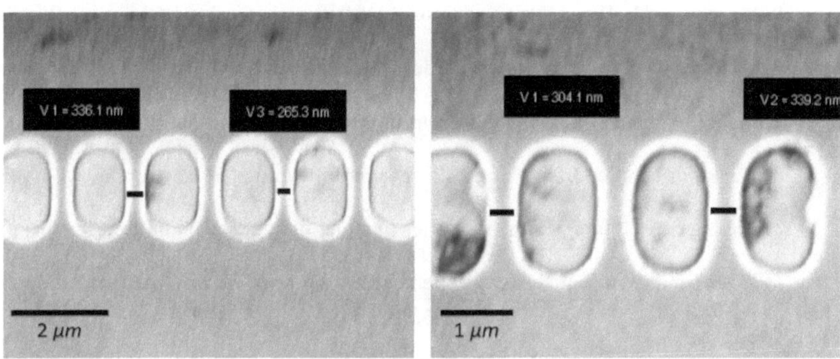

Figure 5.6 Dose and defocus optimization of the fins were carried out and the optimized fin shapes obtained are shown above. The best features were obtained for 1/0.5 microns (fin spacing/fin width) for a dose/defocus of 95mJ cm^{-2}/1

During the initial stages of the process development, a number of $Al_{0.60}Ga_{0.40}N$ samples were used for a comprehensive dose and defocus test. A Heidelberg Maskless Aligner (MLA) was used for the full development of these patters. The photoresist of choice was S1813, due to its high-fidelity pattern reproduction ability and its good etch resistance capabilities. The photoresist (PR) was spin-coated on the samples with a PR thickness of 1.5 microns.

All the dose tests were performed on 1 cm by 1 cm sample which is the typical sample size used in this book. Initially a dose rate of 50 to 150 (steps of 25) of was used together with a defocus ranging from -8 to 8 (steps of 4). Based on these results, another dummy exposure was completed with a smaller dose and defocus range, where the dose was varied from 75 to 115 mJ/cm^2 in steps of 10 mJ/cm^2 and the defocus was varied from 0-2 in steps of 1. The resulting patterns were then etched using a Cl$_2$-based chemistry and the patterns were inspected under a microscope. Figure 5.6 shows the best results from this initial experiment which resulted from a dose rate of 95 mJ/cm^2 and defocus of 1. The thinnest fins resulted from using a fin spacing/fin width of 1/0.5 microns, with a fin width averaging near 300 nm. This is smaller than the light wavelength used for this exposure (375 nm) and is attributed to controlled overexposure of the fins to achieve the desired fin width.

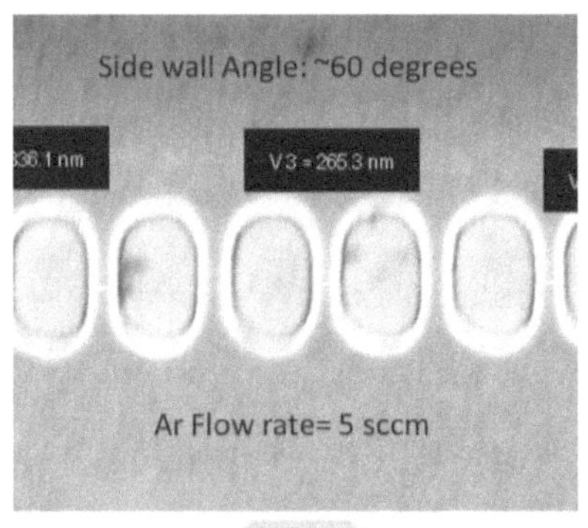

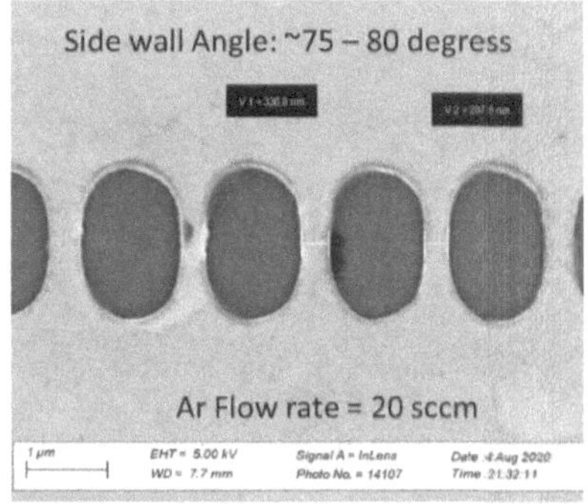

Figure 5.7 Optimization of the side wall angle for the fabricated FinFET structures. Sidewalls, as steep as 80 degrees, were obtained with an Ar flow rate of 20 sccm. The side profile schematics of the fins are illustrated under the SEM images.

For the next step, focus was placed on the optimization of the achieved fin geometry. From the SEM images taken on the initial fins and our previous knowledge on this recipe, it was known that the sidewall angles were not steep and were slanted around 50 to 60 degrees. To increase the steepness of the slope the etch chemistry was modified to increase Ar flow rate which will increase the physical etching component in the etch thus making it steeper. Using an Ar-flow rate of 20 sccm, as opposed to 5 sccm being used during the initial development of the fins, a steep side angle of 75 to 80 degrees was achieved (Figure 5.7).

5.4. COMPARISON OF ELECTRICAL CHARACTERISTICS BETWEEN FINFETS AND PLANAR DEVICES

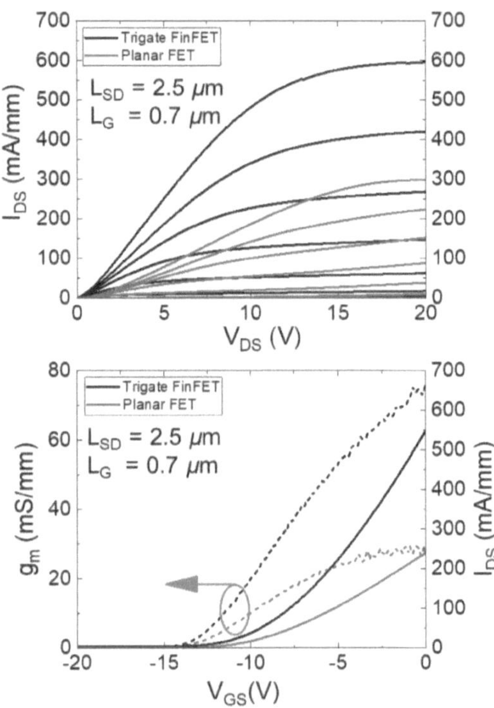

Figure 5.8 Comparison of the planar FET vs FinFET (a) output and (b) transfer characteristics

For this study, a 100 nm thick n-type $Al_{0.60}Ga_{0.40}N$ sample with a doping of 2×10^{18} cm^{-3} was used. This is the same sample that was used to demonstrate device results for the MBE regrown side contacts in Chapter 2. As can be seen from Figure 5.8, significant improvement in the current density is obtained for the FinFET compared to the planar FET device. There is also a significant improvement in the contact resistance of the devices. It is interesting to also note that the contact resistance improvement is most noticeable under low bias conditions.

5.5. SUMMARY

In this chapter, an analysis of the expected improvement in contact resistance by moving from a planar device structure to a FinFET type structure is discussed. An analytical model of the breakdown characteristics of the FinFET devices is also developed. The developed model is then used to a perform a first order analysis of the RF performance of these devices and then compared with planar devices. In the later sections, a process is developed which can be used to fabricate these FinFET like structures, which is then used to fabricate FinFETs on AlGaN channel MESFETs with regrown MBE contact layers. Finally, a comparison in the electrical characteristics of the fabricated FinFETs and planar transistors are provided.

CHAPTER 6:

CONCLUSIONS AND FUTURE WORK

6.1. CONCLUSIONS

This book has studied the design, fabrication and characterization of ultra-wide bandgap $Al_xGa_{1-x}N$ (x > 0.5) based devices for applications in high frequency and high-power electronics in details. These set of materials, despite clear benefits availed by their extremely large critical breakdown electric fields, have some drawbacks that has so far been the bottleneck in the fabrication of high-performance devices. Through this book, novel solutions to many of the long-standing problems have been proposed and subsequently record-breaking devices were demonstrated.

A large portion of this book is dedicated to developing an understanding of the contact formation process in high Al-composition AlGaN. Advanced designs that employ heterostructure engineered contact layers have been thoroughly explored. Two separate approaches for contact formation to MOCVD grown channel AlGaN channel devices have

122

been developed. One method used a novel MBE regrown contact layer to form low resistance contacts to MOCVD grown devices and the other method employed a slow graded in situ MOCVD grown contact layer to establish low resistance contacts to MOCVD grown channel. Besides this, a better understanding of low field mobility in AlGaN channel HEMTs has been developed by the measurement and subsequent modeling of the low field mobility in $Al_{0.70}Ga_{0.30}N/ Al_{0.50}Ga_{0.50}N$ HEMT devices.

A novel solution of improving premature breakdown in lateral devices by the integration of high dielectric constant material has been proposed and demonstrated in Chapter 4. FinFET devices have been proposed and demonstrated as a solution to the further improvement of contact resistance in AlGaN channel devices. An analytical model for understanding the breakdown performance of ultra-wide bandgap FinFET devices have also been developed.

Despite the significant improvement in AlGaN channel device performance that has been demonstrated in these works, there remains room for improvement. In the next few sections, novel approaches are discussed that can further improve AlGaN device performance in the future.

6.2. ONGOING AND FUTURE WORK

6.2.1. ALGAN FINFET MULTI-CHANNEL HEMTS

As discussed in Chapter 3, mobility engineering in AlGaN channel devices by reducing the channel sheet charge density can increase AlGaN channel mobility. To increase the current density in these structures, it is possible to increase the number of parallel conducting channels in the devices which can be realized by using a vertically stacked multi-channel configuration. The combination of the low sheet charge per channel and large number of parallel conducting channels, it is possible to simultaneously achieve high current density and high mobility. These structures have also been discussed in Chapter 5 as a potential solution to reduce effective contact resistance. Due to the benefits that FinFETs enable in high Al-composition AlGaN channel transistors, it would be very interesting to explore these structures in the future. In this book, the process technology to develop such FinFETs has been established. However, further process improvements will still need to be performed including but not limited to fin sidewall angle optimization and reducing fin width to improve gate control and breakdown performance.

6.2.2. N-POLAR HEMTS AND POLFETS

Compared to Gallium polar HEMTs, Nitrogen polar (N- polar) AlGaN channel HEMTs have been shown to have excellent high-power performance at high frequencies due to inherently having a 2DEG distribution closer to the HEMT surface which significantly improves short-channel effects by having a superior aspect ratio. N-polar AlGaN channel POLFETs will have a reduced Al-composition towards the channel surface which would significantly reduce difficulty of contact formation to such devices. Both N-polar compositionally graded POLFETs and HEMTs are thus an area that should be explored thoroughly. Currently, not many such device structures have been explored, due to the difficulty of growing high-quality N-polar AlGaN films. It is expected that as the growth technology improves, more groups and researchers would lean onto using N-polar based AlGaN channel devices due to the obvious benefits they provide.

6.2.3. RADIATION EFFECTS ON ALGaN CHANNEL DEVICES

Some initial work on gamma radiation effects on $Al_{0.60}Ga_{0.40}N$ has been performed. The irradiated samples were 5×10^{18} cm^{-3} doped MESFET samples which were exposed to an accumulated radiation dose of 5.04 Mrads. Electrical measurements were performed pre-radiation and also post-radiation on the same devices. As the plots above demonstrate, no significant or noticeable effects were observed for the radiation type and dosage. However, more studies need to be performed to thoroughly evaluate the radiation hardness of AlGaN.

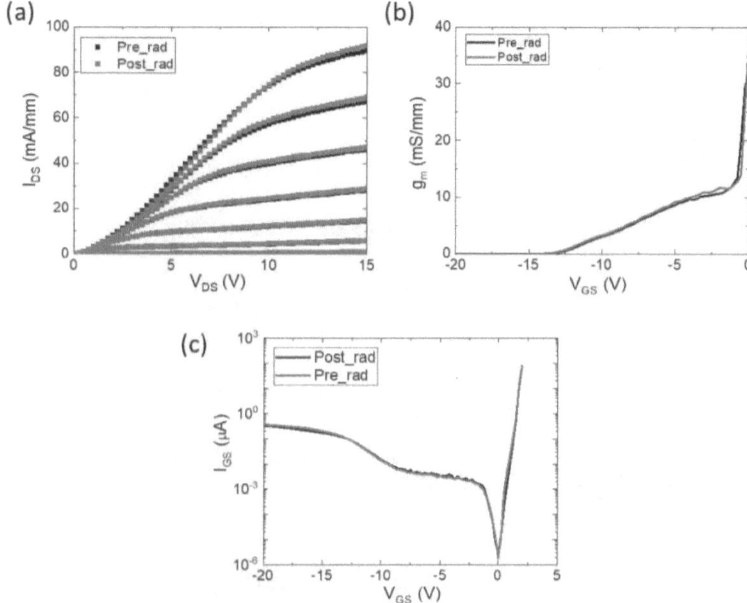

Figure 6.1 Comparison of the (a) I_{DS}-V_{DS}, (b) transonductance and (c) I_{GS}-V_{GS} of the pre-radiation and post-radiation features of a typical gamme irradiated sample is shown above. No significant radiation related effects were observed.

REFERENCES

1. Wu, J., et al., *Unusual properties of the fundamental band gap of InN.* Applied Physics Letters, 2002. **80**(21): p. 3967-3969.

2. Yim, W., et al., *Epitaxially grown AlN and its optical band gap.* Journal of Applied Physics, 1973. **44**(1): p. 292-296.

3. *www.i-micronews.com*. 2020.

4. Hindle, P. *microwavejournal.com*. [cited 2021].

5. *yole.fr*. 2019.

6. Ambacher, O., et al., *Two dimensional electron gases induced by spontaneous and piezoelectric polarization in undoped and doped AlGaN/GaN heterostructures.* Journal of applied physics, 2000. **87**(1): p. 334-344.

7. Wu, Y.-F., et al., *High Al-content AlGaN/GaN MODFETs for ultrahigh performance.* IEEE Electron Device Letters, 1998. **19**(2): p. 50-53.

8. Keller, S., et al., *Gallium nitride based high power heterojunction field effect transistors: Process development and present status at UCSB.* IEEE Transactions on Electron Devices, 2001. **48**(3): p. 552-559.

9. Ambacher, O., et al., *Two-dimensional electron gases induced by spontaneous and piezoelectric polarization charges in N-and Ga-face AlGaN/GaN heterostructures.* Journal of applied physics, 1999. **85**(6): p. 3222-3233.

10. Bhapkar, U.V. and M.S. Shur, *Monte Carlo calculation of velocity-field characteristics of wurtzite GaN.* Journal of Applied Physics, 1997. **82**(4): p. 1649-1655.

11.   Gelmont, B., K. Kim, and M. Shur, *Monte Carlo simulation of electron transport in gallium nitride.* Journal of applied physics, 1993. **74**(3): p. 1818-1821.

12.   Shinohara, K., et al. *Self-aligned-gate GaN-HEMTs with heavily-doped n+-GaN ohmic contacts to 2DEG.* in *2012 International Electron Devices Meeting.* 2012. IEEE.

13.   Yue, Y., et al., *InAlN/AlN/GaN HEMTs with regrown ohmic contacts and f_{T} of 370 GHz.* IEEE Electron Device Letters, 2012. **33**(7): p. 988-990.

14.   Silvaco, I., *ATLAS user's manual.* Santa Clara, CA, Ver, 2011. **5**.

15.   Fang, T., et al., *Effect of optical phonon scattering on the performance of GaN transistors.* IEEE Electron Device Letters, 2012. **33**(5): p. 709-711.

16.   Jena, D., *Polarization induced electron populations in III-V nitride semiconductors: Transport, growth, and device applications.* 2003, Citeseer.

17.   Johnson, E. *Physical limitations on frequency and power parameters of transistors.* in *1958 IRE International Convention Record.* 1966. IEEE.

18.   Xia, Z., *Materials and Device Engineering for High Performance β-Ga2O3-based Electronics.* 2020, The Ohio State University.

19.   Liu, W., et al., *6.2 W/mm and Record 33.8% PAE at 94 GHz from N-polar GaN Deep Recess MIS-HEMTs with ALD Ru Gates.* IEEE Microwave and Wireless Components Letters, 2021.

20.   Romanczyk, B., et al., *W-band power performance of SiN-passivated N-polar GaN deep recess HEMTs.* IEEE Electron Device Letters, 2020. **41**(3): p. 349-352.

21.   Wienecke, S., et al., *N-polar GaN cap MISHEMT with record power density exceeding 6.5 W/mm at 94 GHz.* IEEE Electron Device Letters, 2017. **38**(3): p. 359-362.

22.   Sohel, S.H., et al., *X-band power and linearity performance of compositionally graded AlGaN channel transistors.* IEEE Electron Device Letters, 2018. **39**(12): p. 1884-1887.

23.   Singisetti, U., T. Razzak, and Y. Zhang, *Wide Bandgap Semiconductor Electronics and Devices.* Vol. 63. 2019: World Scientific.

24.   Razzak, T., et al. *Ultra-wide band gap materials for high frequency applications.* in *2018 IEEE MTT-S International Microwave Workshop Series on Advanced Materials and Processes for RF and THz Applications (IMWS-AMP).* 2018. IEEE.

25.   Razzak, T., S. Rajan, and A. Armstrong, *Ultra-wide bandgap AlxGa1-xN channel transistors.* International Journal of High Speed Electronics and Systems, 2019. **28**(01n02): p. 1940009.

26.   Anwar, A., S. Wu, and R.T. Webster, *Temperature dependent transport properties in GaN, Al/sub x/Ga/sub 1-x/N, and In/sub x/Ga/sub 1-x/N semiconductors.* IEEE Transactions on Electron devices, 2001. **48**(3): p. 567-572.

27.   Lie, D.Y., J.C. Mayeda, and J. Lopez. *Highly efficient 5G linear power amplifiers (PA) design challenges.* in *2017 International Symposium on VLSI Design, Automation and Test (VLSI-DAT).* 2017. IEEE.

28.   Razzak, T., et al., *RF operation in graded Al x Ga1− x N (x= 0.65 to 0.82) channel transistors.* Electronics Letters, 2018. **54**(23): p. 1351-1353.

29.   Armstrong, A.M., et al., *Ultra-wide band gap AlGaN polarization-doped field effect transistor.* Japanese Journal of Applied Physics, 2018. **57**(7): p. 074103.

30.   Grundmann, M., *BandEng.* http:/my. ece. ucsb. edu/mgrundmann/bandeng/, 2004.

31.   Baca, A.G., et al., *An AlN/Al0. 85Ga0. 15N high electron mobility transistor.* Applied Physics Letters, 2016. **109**(3): p. 033509.

32.   Baca, A.G., et al. *Al 0.45 Ga 0.55 N/Al 0.30 Ga 0.70 N high electron mobility transistors with Schottky gates and small subthreshold slope factor.* in *2017 75th Annual Device Research Conference (DRC).* 2017. IEEE.

33.   Bajaj, S., et al., *AlGaN channel field effect transistors with graded heterostructure ohmic contacts.* Applied Physics Letters, 2016. **109**(13): p. 133508.

34.   Bajaj, S., et al., *Modeling of high composition AlGaN channel high electron mobility transistors with large threshold voltage.* Applied Physics Letters, 2014. **105**(26): p. 263503.

35.   Douglas, E., et al., *Ohmic contacts to Al-rich AlGaN heterostructures.* physica status solidi (a), 2017. **214**(8): p. 1600842.

36.   Hu, X., et al., *Doped barrier Al 0.65 Ga 0.35 N/Al 0.40 Ga 0.60 N MOSHFET with SiO 2 gate-insulator and Zr-based ohmic contacts.* IEEE Electron Device Letters, 2018. **39**(10): p. 1568-1571.

37.   Muhtadi, S., et al., *Selective area deposited n-Al0. 5Ga0. 5N channel field effect transistors with high solar-blind ultraviolet photo-responsivity.* Applied Physics Letters, 2017. **110**(17): p. 171104.

38. Muhtadi, S., et al., *High electron mobility transistors with Al 0.65 Ga 0.35 N channel layers on thick AlN/sapphire templates.* IEEE Electron Device Letters, 2017. **38**(7): p. 914-917.

39. Nanjo, T., et al., *AlGaN channel HEMT with extremely high breakdown voltage.* IEEE transactions on electron devices, 2013. **60**(3): p. 1046-1053.

40. Nanjo, T., et al., *AlGaN channel HEMTs on AlN buffer layer with sufficiently low off-state drain leakage current.* Electronics letters, 2009. **45**(25): p. 1346-1348.

41. Nanjo, T., et al., *First operation of AlGaN channel high electron mobility transistors.* Applied physics express, 2007. **1**(1): p. 011101.

42. Nanjo, T., et al., *Remarkable breakdown voltage enhancement in AlGaN channel high electron mobility transistors.* Applied Physics Letters, 2008. **92**(26): p. 263502.

43. Raman, A., et al., *AlGaN channel high electron mobility transistors: Device performance and power-switching figure of merit.* Japanese Journal of Applied Physics, 2008. **47**(5R): p. 3359.

44. Xue, H., et al., *Al0. 75Ga0. 25N/Al0. 6Ga0. 4N heterojunction field effect transistor with fT of 40 GHz.* Applied Physics Express, 2019. **12**(6): p. 066502.

45. Yafune, N., et al., *AlN/AlGaN HEMTs on AlN substrate for stable high-temperature operation.* Electronics letters, 2014. **50**(3): p. 211-212.

46. Bajaj, S., *Design and engineering of AlGaN channel-based transistors.* 2018, The Ohio State University.

47. Kaplar, R., et al., *Ultra-wide-bandgap AlGaN power electronic devices.* ECS Journal of Solid State Science and Technology, 2016. **6**(2): p. Q3061.

48. France, R., et al., *Vanadium-based Ohmic contacts to n-Al Ga N in the entire alloy composition.* Applied physics letters, 2007. **90**(6): p. 062115.

49. Mori, K., et al., *Low-ohmic-contact-resistance V-based electrode for n-type AlGaN with high AlN molar fraction.* Japanese Journal of Applied Physics, 2016. **55**(5S): p. 05FL03.

50. Soltani, A., et al., *Development and analysis of low resistance ohmic contact to n-AlGaN/GaN HEMT.* Diamond and related materials, 2007. **16**(2): p. 262-266.

51. Tokuda, H., et al., *High Al composition AlGaN-channel high-electron-mobility transistor on AlN substrate.* Applied physics express, 2010. **3**(12): p. 121003.

52.   Vanko, G., et al., *Nb-Ti/Al/Ni/Au based ohmic contacts to AlGaN/GaN*. Vacuum, 2007. **82**(2): p. 193-196.

53.   Vertiatchikh, A., et al., *Structural properties of alloyed Ti/Al/Ti/Au and Ti/Al/Mo/Au ohmic contacts to AlGaN/GaN*. Solid-State Electronics, 2006. **50**(7-8): p. 1425-1429.

54.   Yafune, N., et al., *Low-resistive ohmic contacts for AlGaN channel high-electron-mobility transistors using Zr/Al/Mo/Au metal stack*. Japanese Journal of Applied Physics, 2011. **50**(10R): p. 100202.

55.   Lee, H.-S., D.S. Lee, and T. Palacios, *AlGaN/GaN high-electron-mobility transistors fabricated through a Au-free technology*. IEEE Electron Device Letters, 2011. **32**(5): p. 623-625.

56.   Wang, L., F.M. Mohammed, and I. Adesida, *Differences in the reaction kinetics and contact formation mechanisms of annealed Ti/ Al/ Mo/ Au Ohmic contacts on n-Ga N and Al Ga N/ Ga N epilayers*. Journal of applied physics, 2007. **101**(1): p. 013702.

57.   Selvanathan, D., et al., *Comparative study of Ti/ Al/ Mo/ Au, Mo/ Al/ Mo/ Au, and V/ Al/ Mo/ Au ohmic contacts to Al Ga N/ Ga N heterostructures*. Journal of Vacuum Science & Technology B: Microelectronics and Nanometer Structures Processing, Measurement, and Phenomena, 2004. **22**(5): p. 2409-2416.

58.   Kobayashi, K.W., D. Denninghoff, and D. Miller. *A Novel 100 MHz-45 GHz GaN HEMT low noise non-gate-terminated distributed amplifier based on a 6-Inch 0.15 μm GaN-SiC mm-wave process technology*. in *2015 IEEE Compound Semiconductor Integrated Circuit Symposium (CSICS)*. 2015. IEEE.

59.   Qiao, D., et al., *Dependence of Ni/AlGaN Schottky barrier height on Al mole fraction*. Journal of Applied Physics, 2000. **87**(2): p. 801-804.

60.   Schroder, D.K., *Semiconductor material and device characterization*. 2015: John Wiley & Sons.

61.   Recht, F., et al., *Nonalloyed ohmic contacts in AlGaN/GaN HEMTs by ion implantation with reduced activation annealing temperature*. IEEE electron device letters, 2006. **27**(4): p. 205-207.

62.   Nagata, N., et al., *Reduction of contact resistance in V-based electrode for high AlN molar fraction n-type AlGaN by using thin SiNx intermediate layer*. physica status solidi c, 2017. **14**(8): p. 1600243.

63.   Bajaj, S., et al., *High Al-content AlGaN transistor with 0.5 A/mm current density and lateral breakdown field exceeding 3.6 MV/cm*. IEEE Electron Device Letters, 2017. **39**(2): p. 256-259.

64.   Razzak, T., et al., *Design of compositionally graded contact layers for MOCVD grown high Al-content AlGaN transistors.* Applied Physics Letters, 2019. **115**(4): p. 043502.

65.   Wood, C. and D. Jena, *Polarization effects in semiconductors: from ab initio theory to device applications.* 2007: Springer Science & Business Media.

66.   Baca, A.G., et al., *RF performance of Al 0.85 Ga 0.15 N/Al 0.70 Ga 0.30 N high electron mobility transistors with 80-nm gates.* IEEE Electron Device Letters, 2018. **40**(1): p. 17-20.

67.   Jena, D., et al., *Realization of wide electron slabs by polarization bulk doping in graded III–V nitride semiconductor alloys.* Applied physics letters, 2002. **81**(23): p. 4395-4397.

68.   Rajan, S., et al., *AlGaN/GaN polarization-doped field-effect transistor for microwave power applications.* Applied physics letters, 2004. **84**(9): p. 1591-1593.

69.   Borisov, B., et al., *Si-doped Al x Ga 1− x N ($0.56 \leqslant \times \leqslant 1$) layers grown by molecular beam epitaxy with ammonia.* Applied physics letters, 2005. **87**(13): p. 132106.

70.   Hwang, J., et al., *Si doping of high-Al-mole fraction Al x Ga 1− x N alloys with rf plasma-induced molecular-beam-epitaxy.* Applied Physics Letters, 2002. **81**(27): p. 5192-5194.

71.   Liang, Y.-H. and E. Towe, *Progress in efficient doping of high aluminum-containing group III-nitrides.* Applied Physics Reviews, 2018. **5**(1): p. 011107.

72.   Mehnke, F., et al., *Highly conductive n-Al x Ga1− x N layers with aluminum mole fractions above 80%.* Applied Physics Letters, 2013. **103**(21): p. 212109.

73.   Pampili, P. and P.J. Parbrook, *Doping of III-nitride materials.* Materials Science in Semiconductor Processing, 2017. **62**: p. 180-191.

74.   Taniyasu, Y., M. Kasu, and N. Kobayashi, *Intentional control of n-type conduction for Si-doped AlN and Al X Ga 1− XN ($0.42 \leqslant x < 1$).* Applied physics letters, 2002. **81**(7): p. 1255-1257.

75.   Klein, B.A., et al., *Planar ohmic contacts to Al0. 45Ga0. 55N/Al0. 3Ga0. 7N high electron mobility transistors.* ECS Journal of Solid State Science and Technology, 2017. **6**(11): p. S3067.

76.   Bastard, G., *Wave mechanics applied to semiconductor heterostructures.* 1988.

77.   Qi, M., et al., *High breakdown single-crystal GaN pn diodes by molecular beam epitaxy.* Applied Physics Letters, 2015. **107**(23): p. 232101.

78.   Nishikawa, A., K. Kumakura, and T. Makimoto, *High critical electric field exceeding 8 MV/cm measured using an AlGaN p–i–n vertical conducting diode on n-SiC substrate*. Japanese journal of applied physics, 2007. **46**(4S): p. 2316.

79.   Xue, H., et al., *All MOCVD grown Al0. 7Ga0. 3N/Al0. 5Ga0. 5N HFET: An approach to make ohmic contacts to Al-rich AlGaN channel transistors*. Solid-State Electronics, 2020. **164**: p. 107696.

80.   Xia, Z., et al., *Design of transistors using high-permittivity materials*. IEEE Transactions on Electron Devices, 2019. **66**(2): p. 896-900.

81.   Muhtadi, S., et al., *High temperature operation of n-AlGaN channel metal semiconductor field effect transistors on low-defect AlN templates*. Applied Physics Letters, 2017. **110**(19): p. 193501.

82.   Lee, N.-Y., et al., *Deposition profile of RF-magnetron-sputtered BaTiO3 thin films*. Japanese journal of applied physics, 1994. **33**(3R): p. 1484.

83.   Shi, Z., Q. Jia, and W. Anderson, *High-performance barium titanate capacitors with double layer structure*. Journal of electronic materials, 1991. **20**(11): p. 939-944.

84.   Li, J., et al., *Giant Electroresistance in Ferroionic Tunnel Junctions*. IScience, 2019. **16**: p. 368-377.

85.   Nalwa, H.S., *Handbook of surfaces and interfaces of materials, five-volume set*. 2001: Elsevier.

86.   Allerman, A., et al., *Al0. 3Ga0. 7N PN diode with breakdown voltage> 1600 V*. Electronics Letters, 2016. **52**(15): p. 1319-1321.

87.   Green, A.J., et al., *3.8-MV/cm Breakdown Strength of MOVPE-Grown Sn-Doped $\beta $-Ga 2 O 3 MOSFETs*. IEEE Electron Device Letters, 2016. **37**(7): p. 902-905.

88.   Nishikawa, A., et al., *High critical electric field of Al x Ga 1− x N p-i-n vertical conducting diodes on n-Si C substrates*. Applied physics letters, 2006. **88**(17): p. 173508.

89.   Volpe, P.-N., et al., *Extreme dielectric strength in boron doped homoepitaxial diamond*. Applied Physics Letters, 2010. **97**(22): p. 223501.

90.   Driche, K., et al., *Characterization of breakdown behavior of diamond Schottky barrier diodes using impact ionization coefficients*. Japanese Journal of Applied Physics, 2017. **56**(4S): p. 04CR12.

91.   Zhe, C.F., *Silicon Carbide: Materials, Processing & Devices*. Vol. 20. 2003: CRC Press.

92.   Lin, P. and H. Leamy, *Tunneling current microscopy*. Applied physics letters, 1983. **42**(8): p. 717-719.

93.   Perkins, C.K., et al., *Demonstration of Fowler–Nordheim Tunneling in Simple Solution-Processed Thin Films*. ACS applied materials & interfaces, 2018. **10**(42): p. 36082-36087.

94.   Xia, Z., et al., *Metal/BaTiO3/β-Ga2O3 dielectric heterojunction diode with 5.7 MV/cm breakdown field*. Applied Physics Letters, 2019. **115**(25): p. 252104.

Process traveler: FinFET Fabrication in a HEMT device

The following is a summary of process steps which can be followed to prepare templates for MBE regrowth. Regrowth of contact layers is discussed on HEMTs (2D channel) and PolFETs (3D channel).

1. Sample cleaning involves a solvent cleaning step:
   - Solvent cleaning:
     - 5 min each of Acetone
     - 5 min each of IPA
     - 5 min each of DI

2. Optical Lithography step to define fin shaped pattern between source and drain regions
   - S1813 photoresist coating:
     - Use a spin-coater to obtain a 1.5 micron thick coating of PR
     - Bake at 105 C for 1 min
     - Expose, develop and inspect pattern

o   2 mins of ashing in O2 plasma to remove PR scum

o   1-2 mins of post development bake to harden PR to prepare for ICP-RIE etch

3. ICP RIE dry etch of ALGaN to form fin shaped structures between source and drain regions

o   Etch Chemistry and flow rate for native oxide etch: BCl_3 = 10 sccm; Pressure = 5 mTorr; RIE/ICP Power = 15 / 60 W; etch period = 2 mins

o   Etch Chemistry and flow rate for AlGaN etch: $BCl_3/Cl_2/Ar$ = 5/50/20 sccm; Pressure = 5 mTorr; RIE/ICP Power = 30 W

o   Wafer carrier backside cooling on. Typically achieved with He flow. Etch is loops of 1 min to avoid excessive heating which can result in non-unform etch rate and PR burn in to the SiO_2

o   Etch rate calibrated to be ~ 45 nm/min for AlGaN 60%

o   Remove PR (hot NMP solvent cleaning)

4. SEM inspection

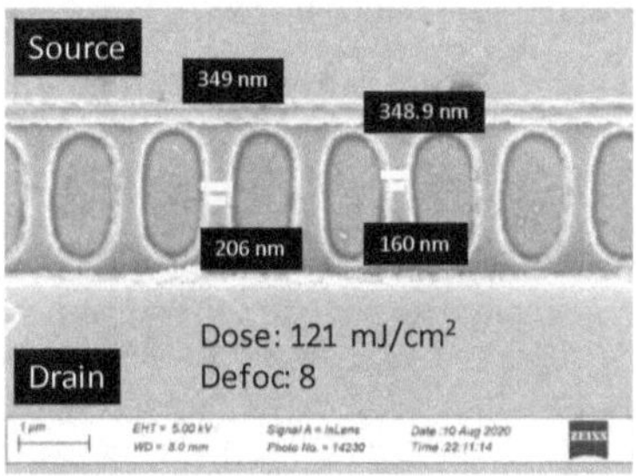

Process traveler: MBE Regrowth using SiO$_2$ mask layer

This section provides a summary of process steps which can be followed to prepare templates for MBE regrowth. Regrowth of contact layers is discussed on HEMTs (2D channel) and PolFETs (3D channel).

1. Sample cleaning involves two cleaning steps:

- Solvent cleaning:
    - 5 min each of Acetone
    - 5 min each of IPA
    - 5 min each of DI
- Optioncal acid cleaning (prior to PECVD SiO$_2$ deposition) to remove native oxide:
    - 1 min conc. HCl or BOE dip followed by 2 mins DI rinse

2. PECVD SiO2

- 500 nm PECVD SiO$_2$ deposition:
    - Pressure = 900 mTorr; N$_2$O/SiH$_4$He = 300/100 sccm
    - RF power = 22 W; substrate temperature = 250°C (or 300°C)
- Ellipsometry:
    - It is recommended to coload a spectator Si sample during the actual deposition. Once deposition is completed an ellipsometer can be used to confirm the thickness of the deposited SiO2 film. Measurement should be may performed both before and after SiO2 deposition to account for the native oxide. Alternatively, a 30 seconds HF dip can be performed to remove the native oxide before deposition

3. Dry etch of SiO$_2$

- A photoresist coating > 1.5 µm is required for the SiO2 etch step
    - S1813 could be used with standard recipe

- Optical lithography for dry etch of SiO_2 for ohmic contact regrowth pattern

- O_2 plasma descum for 2 mins

- ICP-RIE etching:

    o Etch Chemistry and flow rate: $CF4/Ar/O_2$ = 20/5/2 sccm; Pressure = 5 mTorr

    o RIE/ICP Power = 120 / 120 W

    o Wafer carrier backside cooling on. Typically achieved with He flow. Etch is loops of 1 min to avoid excessive heating which can result in non-unform etch rate and PR burn in to the SiO_2

    o Etch rate calibrated to be ~ 68-70 nm/min; recommended etching 6-7 mins

    o Remove PR (hot NMP solvent cleaning)

4. Wet etch SiO_2 to expose AlGaN surface

- Etching calibration prior to the actual etching:

- Etching Condition : BOE (1:10) : DI = 1 : 15, 30 min stirring prior to the calibration

    o 2 mins DI rinse

    o Etch rate calibrated to be ~35 nm/min

    o 2-3 min dip + 2 min DI rinse

This step is to wet etch remaining SiO2 to avoid plasma on the substrate. For PolFET

structures, regrowth may be carried out at this step for a top contact (skip steps 5 and

6). Steps 5 and 6 prepare substrate for side contact

5. Cl2 based (Al)GaN dry etch

- Total Etch 100 nm for standard HEMT (may vary based on structure):

    o BCl3 = 10 [sccm]; RIE/ICP = 15/40 [W], 5 mTorr (1-2 mins) Native oxide removal

- o Cl_2/BCl_3 = 50/5 [sccm], RIE/ICP = 40 V/40 W, Pressure = 5 mTorr

- o Etch rate calibrated to be ~10 nm/min for GaN. Etch rate of (Al)GaN is 10-20% slower and is a function of the Al-composition of the film. Etch rate of SiO_2 ~ 1 nm/min.

6. ~50 nm SiO_2 wet etch extension

- Use the same diluted BOE solution:

    - o 2-3 min dip + 2 min DI rinse

    - o SEM inspection or slow AFM scan to ensure etch extension which is critical to avoid any crevice at interface between regrown contact and channel region

7. MBE regrowth of n^{++} reverse Al-composition graded (Al)GaN

    - o MBE regrowth

    - o AFM / SEM inspection to ensure proper contact between regrown contact layer and channel

8. Wet etch poly-crystalline (Al)GaN

- Backside Indium bonding metal removal:

    - o HCl dip 10 min + 2 min DI rinse

- Heated dil. KOH solution:

    - o ~15% wt; KOH 45%:DI = 1 : 3.5

    - o Solution heated at 75°C using a hot plate

    - o Etch duration calibrated: ~ 5 min @ 75°C

    - o 2 min DI rinse

    - o This step may be done after SiO_2 lift-off to avoid long KOH dip. After lift-off, KOH dip can be done in (1 min dip + 1 min DI rinse + inspection) cycles until remaining poly-crystalline GaN is completely removed.

- SiO_2 lift-off process:

    o   15-30 min BOE (10:1) No Dilution No Ultrasonication

    o   10 min with Ultrasonication

    o   5 min DI rinse with Ultrasonication

- SiO_2 wet etch:

    o   If lift-off is not done, SiO_2 regrowth mask can be removed using BOE

    o   5 min BOE dip + 2 min DI rinse

APPENDIX B

MATLAB CODES FOR THE SIMULATION OF FINFET E-FIELD:

```matlab
% Copyright: Towhidur Razzak, 14th April 2020
% This code calculates the 3D Electric Field profile
% between the gate and drain  for a double gate transistor
%
%    Axis direction:
%
%      (z - bottom of fin to top)
%       |
%       |      / (y - from gate to drain)
%       |    /
%       |  /
%       | /
%       |/
%       |/---------(x - between two gates)
%

clearvars
close all
clc

q=1.6e-19;
eps0=8.854e-14;
epsr=8.9;

Wfin=300e-7;
t=300e-7;%Fin Height
Wby2=Wfin/2; %Half of Fin Width
Ldep=1e-4;
```

```matlab
nst=3e13;%Top 2DEG
n3d=nst/t;
nss=n3d*Wby2;


numPlots=200;

y1=1e-9;
x1=1e-9;

%--------Variables--------------------------
xstart=-Wby2;
xstop=Wby2;
x=linspace(xstart,xstop,numPlots);

ystart=0;
ystop=Ldep;
y=linspace(ystart,ystop,numPlots);

[x,y]=meshgrid(x,y);
%-------------------------------------------


Ex0=((q*nss)/(epsr*eps0)).*(x./Wby2);

% Exa=-((q*n3d)/(4*pi*epsr*eps0))*(((Ldep-
y1).*log(((x+Wby2).^2+(y1-Ldep).^2)./((x-Wby2).^2+(y1-
Ldep).^2)))+(y1.*log(((x+Wby2).^2+y1.^2)./((x-
Wby2).^2+y1.^2))));
% Exb=-((q*n3d)/(4*pi*epsr*eps0)).*(2.*(x-Wby2).*(atan((y1-
Ldep)./(x-Wby2))-atan(y1./(x-Wby2))));
% Exc=-((q*n3d)/(4*pi*epsr*eps0)).*(2.*(x+Wby2).*(atan((y1-
Ldep)./(x+Wby2))-atan(y1./(x+Wby2))));
% Ex_side=Exa-Exb+Exc;
%
% Ex_tot=Ex0+Ex_side;

Exa=((q*n3d)/(4*pi*epsr*eps0))*(((Ldep-
y).*log(((x+Wby2).^2+(y-Ldep).^2)./((x-Wby2).^2+(y-
Ldep).^2)))-((-y).*log(((x+Wby2).^2+(-y).^2)./((x-
Wby2).^2+(-y).^2))));
Exb=((q*n3d)/(4*pi*epsr*eps0)).*(2.*(x-Wby2).*(atan((Ldep-
y)./(x-Wby2))-atan((-y)./(x-Wby2))));
Exc=((q*n3d)/(4*pi*epsr*eps0)).*(2.*(x+Wby2).*(atan((Ldep-
y)./(x+Wby2))-atan((-y)./(x+Wby2))));
```

```matlab
Ex_side=Exa-Exb+Exc;

Ex_tot=Ex0+Ex_side;


Eya=((q*n3d)/(4*pi*epsr*eps0))*(((Wby2-x).*log(((x-
Wby2).^2+y.^2)./((x-Wby2).^2+(y-
Ldep).^2)))+((x+Wby2).*log(((x+Wby2).^2+y.^2)./((x+Wby2).^2
+(y-Ldep).^2))));
Eyb=((q*n3d)/(4*pi*epsr*eps0)).*(2.*(y-Ldep).*(atan((Wby2-
x)./(y-Ldep))-atan((-Wby2-x)./(y-Ldep))));
Eyc=((q*n3d)/(4*pi*epsr*eps0)).*(2.*y.*(atan((Wby2-x)./y)-
atan((-Wby2-x)./y)));
Ey=Eya-Eyb+Eyc;

Eymaxa=((q*n3d)/(4*pi*epsr*eps0))*(((Wby2-x1).*log(((x1-
Wby2)^2+y1.^2)./((x1-Wby2)^2+(y1-Ldep).^2)))-((-Wby2-
x1).*log(((x1+Wby2)^2+y1.^2)./((x1+Wby2)^2+(y1-
Ldep).^2))));
Eymaxb=((q*n3d)/(4*pi*epsr*eps0)).*(2.*(y1-
Ldep).*(atan((Wby2-x1)./(y1-Ldep))-atan((-Wby2-x1)./(y1-
Ldep))));
Eymaxc=((q*n3d)/(4*pi*epsr*eps0)).*(2.*y1.*(atan((Wby2-
x1)./y1)-atan((-Wby2-x1)./y1)));
Eymax=Eymaxa-Eymaxb+Eymaxc;

Ey_y0= Ey+Eymax;

Etot=sqrt((Ex_tot).^2+(Ey_y0).^2);

max(Etot(:))

mesh(x,y,abs(Etot));xlabel('x (cm)');ylabel('y
(cm)');set(gca,'FontSize',20);pbaspect([1 1 1]);
 %mesh(x,y,abs(Ey1));xlabel('Fin Width (cm)');ylabel('Gate
to Drain (cm)');zlabel('Ey
(V/cm)');set(gca,'FontSize',20);pbaspect([1 1 1]);

%Vgd=-Ldep*((q*n3d)/(4*pi*epsr*eps0))*(x-
Wfin)*log(((x+Wfin)^2+Ldep^2)/((x+Wfin)^2))

%mesh(x,y,Etot);xlabel('Fin Width');ylabel('Gate to
Drain');zlabel('Electric Field');

%subplot(2,2,1);
%plot(y,Ey1);
```

```matlab
% subplot(2,2,2);
% plot(y,Ex);
%
% subplot(2,2,3)
% plot(y,Etot);
```

MATLAB CODES FOR THE SIMULATION OF FINFET E-FIELD:

```matlab
% Copyright: Towhidur Razzak, 14th April 2020
% This code calculates the 3D Electric Field profile
% between the gate and drain  for a double gate transistor
%
%    Axis direction:
%
%      (z - bottom of fin to top)
%      |
%      |       / (y - from gate to drain)
%      |     /
%      |   /
%      | /
%      |/---------(x - between two gates)
%
%\\\\\\\\\\\\\\\\\\\\\\\\\\\\\\\\\\\\\\\\\\\\\\\\\\\\\\\\\\\\\\\\
%\\\\\\\\\\\\\\\\\\\\\\\\\\\\\\\\\\\\\\\\\\\\\\\\\\\\\\\\\\\\\\\\
%\\\\\\\\\\\\\\\\\\\\\\\\\\\\\\\\\\\\\\\\\\\\\\\\\\\\\\\\\\\\\\\\

clearvars
close all
clc

q=1.6e-19;
eps0=8.854e-14;
epsr=8.9;

Wfin=300e-7;
t=300e-7;               %Fin Height
Wby2=Wfin/2;            %Half of Fin Width
Ldep=1e-4;

nst=3e13;               %Top 2DEG
```

```matlab
n3d=nst/t;
nss=n3d*Wby2;

numPoints=1000;

x=0;
x1=1e-9;
y1=1e-9;

%--------Variables------------------------
ystart=0;
ystop=Ldep;
y=linspace(ystart,ystop,numPoints);
%yinc=(ystop-ystart)/numPoints;
%y=ystart:yinc:ystop;
%-----------------------------------------

Eya=((q*n3d)/(4*pi*epsr*eps0))*(((Wby2-x).*log(((x-
Wby2)^2+y.^2)./((x-Wby2)^2+(y-Ldep).^2)))-((-Wby2-
x).*log(((x+Wby2)^2+y.^2)./((x+Wby2)^2+(y-Ldep).^2))));
Eyb=((q*n3d)/(4*pi*epsr*eps0)).*(2.*(y-Ldep).*(atan((Wby2-
x)./(y-Ldep))-atan((-Wby2-x)./(y-Ldep))));
Eyc=((q*n3d)/(4*pi*epsr*eps0)).*(2.*y.*(atan((Wby2-x)./y)-
atan((-Wby2-x)./y)));

Ey=Eya-Eyb+Eyc;

Eymaxa=((q*n3d)/(4*pi*epsr*eps0))*(((Wby2-x1).*log(((x1-
Wby2)^2+y1.^2)./((x1-Wby2)^2+(y1-Ldep).^2)))-((-Wby2-
x1).*log(((x1+Wby2)^2+y1.^2)./((x1+Wby2)^2+(y1-
Ldep).^2))));
Eymaxb=((q*n3d)/(4*pi*epsr*eps0)).*(2.*(y1-
Ldep).*(atan((Wby2-x1)./(y1-Ldep))-atan((-Wby2-x1)./(y1-
Ldep))));
Eymaxc=((q*n3d)/(4*pi*epsr*eps0)).*(2.*y1.*(atan((Wby2-
x1)./y1)-atan((-Wby2-x1)./y1)));
Eymax=Eymaxa-Eymaxb+Eymaxc;

Ey1=Ey+Eymax;

% figure(1)
 plot(y,abs(Ey1));set(gca,'FontSize',20);pbaspect([1 1 1]);
Vgd=abs(trapz(y,Ey1))

peak = max(abs(Ey1));
```

```matlab
%
% Ex=(q*nss)/(epsr*eps0);
%
% Etot=(Ex^2+Ey1.^2).^(1/2);
%
% Vgd=-Ldep*((q*n3d)/(4*pi*epsr*eps0))*(x-
Wfin)*log(((x+Wfin)^2+Ldep^2)/((x+Wfin)^2))

%subplot(2,2,1);
% plot(y,Eya);
% hold on
% plot(y,Eyb);
% plot(y,Eyc);


% subplot(2,2,2);
% plot(y,Ex);
%
% subplot(2,2,3)
% plot(y,Ey);
```

SILVACO CODES FOR THE SIMULATION OF BTO/ALGAN DIODES:

```
{
## Code to simulate Ga Polar GaN/AlGaN HEMTs using Optical
Phonon Limited Velocity Saturation Model
## Written by - P.S.Park, O.F.Shoron, S.Bajaj

go atlas
#simflags="-P 64"

Set
LABEL="BTO_HEMT_simulation_ns=1.5e13_eps_r=10_LAC=0.20_0416
"

## Setting bias voltages

Set VgStart=1
Set Vd1=10

#polarization: 1 for Ga Polar, -1 for N Polar, 0 for Non
Polar

Set pol=1

Set permit=1000

Set Vf=-200

# Lateral device dimensions
Set Lg=.25
Set Lgs=.1
Set Lgd=0.20
# Lc: Source and Drain Contact width
Set Lc=0.2

# variables to define device structure
Set Ls0=-$Lg/2
#Set Ls1=-$Lg/2-$Lgs
#Set Ls2=$Ls1-$Lc

Set Ls1=$Ls0
Set Ls2=$Ls0


Set Ld0=$Lg/2
Set Ld1=$Lg/2+$Lgd
```

```
Set Ld2=$Ld1+$Lc


# Vertical layer thickness

Set tAir=1
#tBar1: AlGaN Barrier
Set tBar1=25e-3
#tBar2: AlN Interlayer
Set tBar2=0
#tCh: Channel thickness
Set tCh=30e-3
#tSub: GaN Substrate thickness
Set tsub=1


## Metal thickness
Set tOhmic=0.1
Set tGate=0.1


# variables to define device structure
set t1=$tBar1
set t2=$t1+$tBar2
set t3=$t2+$tCh
set t4=$t3+$tsub


# Mesh

mesh space.mult=1

## Set x-dir meshes
#x.m l=$Ls2 s=0.1
#x.m l=$Ls1-5e-3 s=0.02
#x.m l=$Ls1 s=20e-3
#x.m l=$Ls0-50e-3 s=10e-3
x.m l=$Ls0 s=10e-3
x.m l=0 s=10e-3
x.m l=$Ld0 s=5e-3
x.m l=$Ld0+0.1 s=10e-3
x.m l=$Ld1 s=20e-3
x.m l=$Ld1+5e-3 s=0.02
x.m l=$Ld2 s=0.1

## Set y-dir meshes
```

```
y.m l=-$tAir s=0.1
y.m l=-0.2 s=0.1
y.m l=-($tGate+0.05) s=0.1
y.m l=-$tGate s=0.01
#y.m l=-0.0005 s=0.005
#y.m l=-0.0002 s=0.2e-3
y.m l=-0.025 s=2e-3
y.m l=0 s=2e-3
#y.m l=0.0002 s=0.2e-3
#y.m l=0.0005 s=0.002
y.m l=0.01 s=2e-3
y.m l=$t1-0.6e-3 s=1e-3
y.m l=$t1-0.3e-3 s=0.2e-3
y.m l=$t1 s=0.2e-3
y.m l=$t2 s=0.5e-3
y.m l=$t2+10e-3 s=1e-3
y.m l=$t2+20e-3 s=4e-3
y.m l=$t3-0.5e-3 s=1e-3
y.m l=$t3 s=0.2e-3
y.m l=$t3+0.3e-3 s=0.2e-3
y.m l=$t3+0.6e-3 s=3e-3
y.m l=$t3+10e-3 s=10e-3
y.m l=$t3+20e-3 s=0.1
y.m l=$t4 s=0.1

# Structure
region num=1 material=Air x.min=$Ls2 x.max=$Ld2 y.min=-
$tAir y.max=0
region num=2 material=Air x.min=$Ls2 x.max=$Ld2 y.min=-0.2
y.max=0
region num=9 user.material=BTO x.min=$Ls1 x.max=$Ld1
y.min=-0.025 y.max=0
region num=3 material=AlGaN x.min=$Ls1 x.max=$Ld1 y.min=0
y.max=$t1 donor=1e15 x.comp=0.28
region num=4 material=AlGaN x.min=$Ls1 x.max=$Ld1 y.min=$t1
y.max=$t2 donor=1e15 x.comp=0.28
region num=5 material=GaN x.min=$Ls1 x.max=$Ld1 y.min=$t2
y.max=$t3 donor=1e15

region num=6 material=AlGaN x.min=$Ls2 x.max=$Ld2 y.min=$t3
y.max=$t4 donor=1e14 x.comp=0.04

#region num=6 material=AlGaN x.min=$Ls2 x.max=$Ld2
y.min=$t3 y.max=$t4 insulator x.comp=0.04
#Highly doped GaN region for Ohmic Contacts
```

```
#region num=7 material=GaN x.min=$Ls2 x.max=$Ls1 y.min=0
y.max=$t3 donor=1e21
#region num=8 material=GaN x.min=$Ld1 x.max=$Ld2 y.min=0
y.max=$t3 donor=1e21


##Electrodes
#Source
#elec num=1 name=source x.min=$Ls2 x.max=$Ls1 y.min=-
$tOhmic y.max=$t3
#Drain
elec num=1 name=drain x.min=$Ld1 x.max=$Ld2 y.min=-$tOhmic
y.max=$t3
#Gate
elec num=2 name=gate x.min=$Ls0 x.max=$Ld0 y.min=-$tGate
y.max=-0.025


## Material
material kp.set2 pol.set2
#material material=AlGaN affinity=3.78625 eg300=3.98875
psp=-0.0425 alattice=3.17 c13=104.25 c33=397 e31=-0.5175
e33=0.9125 taun0=1e-9 taup0=1e-9 copt=1.1e-8 augn=1.0e-34
augp=1.0e-34 mup=1 vsatp=1 F.TOFIMUN=OP_gan5.lib
#material material=GaN affinity=4.1 eg300=3.42 psp=-0.029
alattice=3.189 c13=103 c33=405 e31=-0.49 e33=0.73 taun0=1e-
9 taup0=1e-9 copt=1.1e-8 augn=1.0e-34 augp=1.0e-34 mup=1
vsatp=1 F.TOFIMUN=OP_gan5.lib
#OP_gan5.lib is the code to include velocity model
material material=BTO user.group=semiconductor
user.default=ZnO EG300=3.25 Mun=8 affinity=4.1
permittivity=$permit M.DSN=1.8
material material=AlGaN F.TOFIMUN=OP_gan5.lib
material material=GaN F.TOFIMUN=OP_gan5.lib

doping x.min=$Ls1 x.max=$Ld1 y.min=$t2-0.2e-3 y.max=$t2
donor conc=6.5e20 uniform
#doping x.min=$Ls1 x.max=$Ld1 y.min=0 y.max=0.2e-3 accep
conc=6e20 uniform
doping x.min=$Ls1 x.max=$Ld1 y.min=$t3 y.max=$t3+0.2e-3
accep conc=1.067e20 uniform
```

```
##Contacts
#contact name=source resistance=200 workfunc=3.8
contact name=drain resistance=200 workfunc=3.8
contact name=gate workfunc=5.40 resistance=400

#Interface trap densities
#inttrap s.i acceptor e.level=1.6 density=5e13 degen=4
sign=2.84e-15 sigp=2.84e-14 x.min=$Ls1 x.max=$Ls0 y.min=0
y.max=0
#inttrap s.i acceptor e.level=1.6 density=5e13 degen=4
sign=2.84e-15 sigp=2.84e-14 x.min=$Ld0 x.max=$Ld1 y.min=0
y.max=0

# Bulk Traps
trap acceptor e.level=3.28 density=4e16 degen=4 sign=2.84e-
15 sigp=2.84e-14 x.min=$Ls2 x.max=$Ld2 y.min=$t3 y.max=$t4

models srh fermi print
#models schro dd_ms sp.geo=1dy print

#models region=9 ferro
#material region=9 ferro.ps=0.5e-15 ferro.pr=0.4e-20
ferro.ec=1.0 ferro.eps=1
#material region=9 F.FERRO=STOpermittivity3.lib

# Solution Method
method newton carr=2 itlim=50 trap maxtrap=20 print
#method gummel newton trap


# Output
output con.band val.band charge polar.charge flowlines
e.field e.velocity ex.velocity ey.velocity e.mobility
e.temp band.par J.diff J.drift permittivity


################################################################
################################## Initial solve

solve init
save outf=init.str
tonyplot init.str
```

```
#quit
###############################################################
################################## Id-Vg

#solve vstep=0.1 vfinal=10 name=drain

#solve vstep=1 vfinal=60 name=drain
#save outf=(60)V_1.str master

#log outf=IdVg_1.log


solve name=gate vgate=0 vstep=-0.5 vfinal=-5.5
save outf=$LABEL(-5.5)V.str master

solve name=gate vstep=-1 vfinal=-8
save outf=$LABEL(-8)V.str master

#solve name=gate vstep=-1 vfinal=-20
#save outf=$LABEL(-20)V.str master

#solve name=gate vstep=-1 vfinal=-130
#save outf=$LABEL(-130)V.str master

#solve name=gate vstep=-1 vfinal=-140
#save outf=$LABEL(-140)V.str master

solve name=gate vstep=-1 vfinal=$Vf
save outf=$LABEL($Vf)V.str master
log off

tonyplot $LABEL($Vf)V.str
quit

}
```